内隐性别刻板印象

理论与实证

顾吉有　著

中国社会科学出版社

图书在版编目（CIP）数据

内隐性别刻板印象：理论与实证／顾吉有著.

北京：中国社会科学出版社，2024.8. —— ISBN 978 - 7 - 5227 - 4205 - 2

Ⅰ．B844

中国国家版本馆 CIP 数据核字第 20244L0C34 号

出 版 人　赵剑英
责任编辑　姜雅雯
责任校对　苏　颖
责任印制　郝美娜

出　　版　中国社会科学出版社
社　　址　北京鼓楼西大街甲 158 号
邮　　编　100720
网　　址　http：//www.csspw.cn
发 行 部　010 - 84083685
门 市 部　010 - 84029450
经　　销　新华书店及其他书店

印　　刷　北京君升印刷有限公司
装　　订　廊坊市广阳区广增装订厂
版　　次　2024 年 8 月第 1 版
印　　次　2024 年 8 月第 1 次印刷

开　　本　710×1000　1/16
印　　张　11.25
插　　页　2
字　　数　141 千字
定　　价　68.00 元

目　　录

前　　言

　　刻板印象是一种较为常见的认知现象。从本质上说，刻板印象的核心是一种对人的分类，这种分类有利于我们快速地认识他人，从而节省自身的认知资源。在生活中，人们对性别刻板印象的使用非常普遍。在学科、职业、外貌和人格特质等领域都体现着性别刻板印象的影响。但是，采用刻板印象对人进行分类的准确性存在争议，有时候刻板印象的内容甚至存在错误，因此，正确认识刻板印象的特征、心理机制、影响因素，认识到刻板印象的影响范围，从而在这些领域尽力避免刻板印象及其负面影响，是心理学和社会学领域重点关注的话题。

　　基于上述考虑，本书首先对性别刻板印象的基本概念、相关特征、心理机制和研究方法进行阐述。其次，针对刻板印象带来的负面影响——刻板印象威胁效应及外溢效应展开讨论，以期能为减少刻板印象对个体或群体的消极影响提供理论和实验证据。最后，本书提供了关于儿童性别刻板印象发展的理论和实证研究，以及情绪对性别刻板印象影响的实证研究。

第　一　章

性别刻板印象概述

第一节　刻板印象与内隐刻板印象的概念与特征

一　刻板印象

刻板印象（stereotype）是人类社会交往活动中一种普遍的认知现象。它的出现是人类进化的必然产物，也是人类进一步认识社会与群体的一种重要的心理基础。从提出之日起，刻板印象便受到心理学家的广泛关注。从词源学的角度看，"stereotype"本身由两个希腊单词——"stereos"和"tupos"——组成。其中，"stereos"意为"坚固的""坚硬的"，"typos"意为记号、压痕、模型。也就是说，词源学中的"stereotype"是"坚固的模型"之意。而在印刷业中，"stereotype"被称作"铅板"。由此可以看出，不管是在词源学还是印刷业中，"stereotype"均有坚固的意思。

心理学中关于"刻板印象"的研究始于20世纪20年代。自1922年新闻工作者李普曼（Lippmann）在其著作《公众舆论》中提出刻板印象这一概念至今，有关刻板印象的研究已有90多年的历史。最初，Lippmann发现，偏见（preconception）能够影响一个人对他人和事物的知觉，并用印刷业的术语"stereotype"来指代这种现象。此后，众多心理学工作者对刻板印象的概念提出了不同的理解。比如，Allport 认为，"刻板印象是与某个类别相联系的一种夸大的观念，它的作用是证明与那个类别有关的行为"[1]；Mackie 认为，"刻板印象是人们的观念，即有关刻画一种社会类别的高度一致的属性的观念"[2]；Hamilton 和 Trolier 对刻板印象的定义是："包括知觉者对某个群体的知识、观念和预期的一种认知结构"[3]；Rudman 等人认为，"刻板印象是一个假设的认知结构，包含推理上的关系的结构集合，这一关系与个人对一个社会类别的态度相连"[4]；Hamilton 和 Sherman 则将刻板印象描述成："对与一个社会群体相联系的思想、事实和想象的认知表征"[5]。我国学者连淑芳对刻板印象的定义是：关于特定群体的特质、属性和行为的一组观念，或者说是对与社会群体及其成员相联系的特质或属性的认知表征。[6]

　　尽管国内外学者对于刻板印象的定义不尽相同，但从这些定义中我们仍可以看出：一方面，刻板印象以非常明显的自然特征区别不同的群体，比如性别、种族等，存在刻板印象的个体会把一系列特征归为某一群体的个体所有，并且认为某一群体的成员

---

[1]　Allport G. W. , *The Nature of Prejudice*, Reading, MA：Addison-Wesley, 1954.

[2]　Mackie A. , "The chemical basis of food detection in the lobster Homarus gammarus", *Marine Biology*, Vol. 21, No. 2, March 1973, pp. 103 – 108.

[3]　Hamilton D. L. , & Trolier T. K. , "Stereotypes and stereotyping：An overview of the cognitive approach", in J. F. Dovidio & S. L. Gaertner（Eds.）, *Prejudice, Discrimination, and Racism*, New York：Academic Press, 1986.

[4]　Rudman L. A. , Ashmore R. D. , & Gary M. L. , "'Unlearning' automatic biases：The malleability of implicit stereotypes and prejudice", *Journal of Personality and Social Psychology*, Vol. 81, April 2001, pp. 856 – 868.

[5]　Hamilton D. L. , & Sherman J. W. , "Stereotypes", in R. S. Wyer, Jr. , & T. K. Srull（Eds.）, *Handbook of Social Cognition*（2nd ed.）, Hillsdale, NJ：Erlbaum, Vol. 2, 1994, pp. 1 – 68.

[6]　连淑芳:《刻板印象的自动过程研究新进展》,《心理科学》2003 年第 1 期。

均具有该群体的全部特征；另一方面，刻板印象是社会印象的一种表现形式，即一种固定的印象。

二 内隐刻板印象

20世纪90年代以前，由于理论水平的限制，人们普遍认为，刻板印象的信息加工是个体能意识到的，因此，那时关于刻板印象的研究均局限于对其外显过程的考察。后来，Greenwald 根据社会信息加工的意识与无意识性，将社会认知划分为外显和内隐两部分，并将内隐社会认知（implicit social cognition）定义为：在社会认知的过程中，虽然个体不能回忆某一过去的经验，但这一经验却潜在地对个体的行为和判断造成影响。[①] 随后，Greenwald 和 Banaji 又将这一概念拓展到刻板印象领域中，提出内隐刻板印象（implicit stereotype）的概念。[②] 内隐刻板印象是指：过去的经验不能被个体内省地获得（或不能准确地识别），但这些经历却影响着对某一群体成员特征的评价。这个概念的提出对社会认知领域的理论更新和方法进步有重要的作用。首先，它认可了刻板印象能内隐地对个体行为和判断造成影响，因而为后续研究将刻板印象的有意识加工和无意识加工相分离提供了理论依据；其次，这个概念本身是对过去的刻板印象领域研究方法的批判，它意味着以往研究方法的局限性，促使研究者采用更为间接的测量方法来考察刻板印象中个体意识不到的部分。

---

① Greenwald A. G. , "What cognitive representations underlie attitudes?" *Bulletin of the Psychonomic Society*, Vol. 28, No. 2, May 1990, pp. 254 – 260.

② Greenwald A. G. & Banaji M. R. , "Implicit Social Cognition. Attitudes, Self-Esteem, and stereotypes", *Psychological Review*, Vol. 102, No. 1, July 1995, pp. 4 – 27.

第二节　内隐刻板印象的研究方法

　　研究方法一直制约着研究领域的发展，社会认知领域也不例外。自研究伊始，大量心理学工作者采用不同的方式来测量刻板印象，从早期的自我报告和行为测量，到现在的神经学测量。技术的更迭代表了研究的进展，不同的测量方法表明刻板印象研究领域从单纯考察个体的行为，发展到了考察个体行为背后的生理神经因素。

　　过去主要采用的方法是自我报告，包括思想列表（thought listings）[1]、特质核对（traits check-offs）[2]、概率判断（probability judgements）[3] 以及李克特量表（Likert scales）。但是，由于被试对量表含义的主观认识不同，上述这些方法可能会扭曲个体对群体的判断。[4][5] 此外，非反应性、间接性或非侵扰性的行为测量[6]，

---

① Stangor C., Sullivan L. A., & Ford T. E., "Affective and cognitive determinants of prejudice", *Social Cognition*, Vol. 9, June 1991, pp. 359 – 380.

② Katz D., & Braly K. W., "Racial stereotypes of one hundred college students", *Journal of Abnormal and Social Psychology*, Vol. 28, May 1933, pp. 280 – 290.

③ McCauley C., & Stitt C. L., "An individual and quantitative measure of stereotypes", *Journal of Personality and Social Psychology*, No. 36, October 1978, pp. 929 – 940.

④ Biernat M., & Fuegen K., "Shifting standards and the evaluation of competence: Complexity in gender-based judgment and decision making", *Journal of Social Issues*, Vol. 57, No. 4, July 2001, pp. 707 – 724.

⑤ Collins E. C., Crandall C. S., & Biernat M., "Stereotypes and implicit social comparison: Shifts in comparison-group focus", *Journal of Experimental Social Psychology*, Vol. 42, No. 4, May 2006, pp. 452 – 459.

⑥ Crosby F., Bromley S., & Saxe L., "Recent unobtrusive studies of black and white discrimination and prejudice: A literature review", *Psychological Bulletin*, No. 87, June 1980, pp. 546 – 563.

如座位距离（sitting distance）[①] 和内隐反应时测量[②③④]也很普遍。但这些研究方法对于实验设计的要求相对较高，因此在实施实验的过程中需要更加谨慎。随着技术的不断进步，从事件相关电位技术到现在的功能性核磁共振技术，研究者开始越来越多地利用脑成像原理对刻板印象进行考察，但是，研究者仍无法确定的是哪一种技术比其他技术更能达到预期效果[⑤]。

一　社会认知领域的测量方法概述

在有关刻板印象或偏见的研究领域，往往可能会涉及有关自我表述的一些社会影响问题。也就是说，研究者认为，如果让被试直接回答和刻板印象或偏见有关的问题，他们并不会真实表达自己的观点，所以，研究者设计了间接测量的方法，他们认为，间接测量能更有效地考察被试的真实想法或态度。事实上，在社会认知领域中，研究者将内隐测量方法沿用至今，尤其是在种族主义和性别歧视方面，一些最重要、最有用的理论都是基于这样

①　Macrae C. N. , Milne A. B. , & Bodenhausen G. V. , "Stereotypes as energy-saving devices: A peek inside the cognitive toolbox", *Journal of Personality and Social Psychology* , No. 66 , February 1994 , pp. 37 – 47.

②　Banaji M. R. , & Hardin C. D. , "Automatic stereotyping", *Psychological Science* , No. 7 , October 1996 , pp. 136 – 141.

③　Cunningham W. , Preacher K. , & Banaji M. , "Implicit attitude measures: Consistency, stability, and convergent validity", *Psychological Science* , No. 12 , May 2001 , pp. 163 – 170.

④　Dasgupta N. , McGhee D. E. , Greenwald A. G. , & Banaji M. R. , "Automatic preference for White Americans: Eliminating the familiarity explanation", *Journal of Experimental Social Psychology* , Vol. 36 , No. 3 , September 2000 , pp. 316 – 328.

⑤　Kubota J. T. , Banaji M. R. , & Phelps E. A. , "The neuroscience of race", *Nature Neuroscience* , Vol. 15 , No. 7 , July 2012 , pp. 940 – 948.

一种观点，即我们的偏见比我们愿意向自己或他人展示的要多[1][2][3]，而且当这些偏见可以被其他外部借口掩盖时，人们会更多地表达这些偏见[4]。

二 内隐刻板印象的测量方法

如前文所述，早期的研究多采用主观报告法，也就是说，被试在问卷或访谈中回答研究人员有关刻板印象的问题。试想，如果我们自己是被试，我们是否愿意在问卷上承认，自己认为男性是勇敢的、女性是温柔的？一旦我们承认自己有这样的想法，是否会导致别人对我们的其他看法？是否会因此招致消极评价呢？如果我们不愿意在问卷上承认，那么内隐测量法可以避开这些直接问题，更准确地揭示我们的潜在态度。

内隐测量的理论源于社会认知理论[5][6]，这个理论强调自发的

---

① Crandall C. S., & Eshleman A., "A justification—suppression model of the expression and experience of prejudice", *Psychological Bulletin*, Vol. 129, No. 3, July 2003, pp. 414 - 446.

② Gaertner S. L., & Dovidio J. F., "Racism among the well intentioned", in E. Clausen & J. Bermingham (Eds.), *Pluralism, Racism and Public Policy: The Search for Equality*, Boston, MA: G. K. Hall, 1981, pp. 208 - 222.

③ Gaertner S. L., & Dovidio J. F., "The aversive form of racism", In S. L. Gaertner & J. F. Dovidio (Eds.), *Prejudice, Discrimination and Racism*, Orlando, FL: Academic, 1986, pp. 1 - 34.

④ Zuwerink J., Devine P., Monteith J., & Cook D., "Prejudice toward blacks: With and without compunction?" *Basic and Applied Social Psychology*, No. 18, October 1996, pp. 131 - 150.

⑤ Gawronski B., & Payne B. K. (Eds.), *Handbook of Implicit Social Cognition: Measurement, Theory, and Applications*, Guilford Press, 2010.

⑥ Greenwald A. G. & Banaji M. R., "Implicit Social Cognition. Attitudes, Self-Esteem, and stereotypes", *Psychological Review*, Vol. 102, No. 1, 1995, pp. 4 - 27.

和无意识的认知过程对判断和行为的作用①，而内隐测量法的开发就是为了挖掘这些认知过程。最近的评价理论认为，简单联想和命题信念之间存在区别，内隐测量法能够较为敏感地测量出简单联想的效应。②③ 另有研究表明，许多内隐测量法对研究对象的情感反应比对其认知反应更敏感。④ 因此，内隐测量法评估的态度往往具有基于对研究对象更多情感联想的自动属性，它们提供了一个态度指标，而无须直接询问。

也就是说，鉴于直接测量已经不能够满足内隐社会认知中的研究需要，研究者逐渐开发出各种间接测量的方法，为内隐社会认知领域的研究提供技术支持。比如，投射测量、加工分离程序和刻板解释偏差（SEB）等，这些方法均被内隐社会认知领域所广泛使用。值得注意的是，基于反应时范式的测量方法，由于其采用的反应时指标能更加客观地揭示内隐社会认知的加工过程，因而得到更多研究者的认同。⑤ 以下将简单介绍几种常见的以反应时为指标的内隐刻板印象研究方法。

---

① Schacter D. L., Chiu C. Y. P., & Ochsner K. N., "Implicit memory: A selective review", *Annual Review of Neuroscience*, No. 16, April 1993, pp. 159 – 182.

② Gawronski B., & Bodenhausen G. V., "The associative-propositional evaluation model: Theory, evidence, and open questions", *Advances in Experimental Social Psychology*, No. 44, April 2011, p. 59.

③ Rydell R. J., & McConnell A. R., "Understanding implicit and explicit attitude change: A systems of reasoning analysis", *Journal of Personality and Social Psychology*, No. 91, May 2006, p. 995.

④ Smith C. T., & Nosek B. A., "Affective focus increases the concordance between implicit and explicit attitudes", *Social Cognition*, No. 42, June 2011, pp. 300 – 313.

⑤ De Houwer J., "Comparing measures of attitudes at the functional and procedural level: Analysis and implications", in R. E. Petty, R. H. Fazio, & P. Briñol (Eds.), *Attitudes: Insights from the New Implicit Measures*, Mahwah, NJ: Erlbaum, 2009, pp. 361 – 390.

（一）词汇决定任务（Lexical Decision Test，LDT）

在词汇决定任务中，被试要判断主试向其呈现的字符串是否为真词。比如，在 Gardener 和 Mclaughlin 的实验中[①]，向被试呈现一对字符串，当这两个字符串都是真词时，如果二者相互联系（如，一个词为"dog"，另一个词为"cat"），那么被试判断的速度要快于二者没有联系的情况（如，一个词为"dog"，另一个词为"fat"）。内隐刻板印象领域的研究者对此范式进行了改进。比如，在 Kawakami 等人的研究中，被试被随机分为两组。一组被试对老年人的照片进行分类，以激活其老年刻板印象；另一组被试则不参与这项任务。随后向被试呈现一系列的字符串，让其判断每个字符串是否为真词。结果发现，进行图片分类任务的被试，在后面的词汇决定任务中将更容易出现错误判断，并且，正确反应的反应时要长于没有进行分类任务的被试。

（二）内隐联想测验（Implicit Association Test，IAT）

内隐联想测验由 Greenwald，McGhee 和 Schwartz 提出，它以神经网络模型为基础。[②] 神经网络模型认为，信息被存储在一系列神经联系的结点上，这些结点是按照语义关系分层组织起来的，在语义上联结较近的两个结点之间的距离也较小。因此，两类概念在这种神经联系上的距离代表二者之间在语义上的联系：

[①]　Gardener S. L. , & McLaughlin J. P. , "Racial stereotypes: Associaltions and ascriptions of positive and negative characteristics", *Social Psychology Quarterly*, No. 46, May 1983, pp. 23 – 30.

[②]　Greenwald A. G. , McGhee D. E. , & Schwartz J. L. K. , "Measuring individual differences in implicit cognition: The implicit association test", *Journal of Personality and Social Psychology*, Vol. 74, No. 6, April 1998, pp. 1464 – 1480.

距离越小，二者的联系越密切；相反，距离越大，二者的联系越小[①]。Greenwald 等人[②]将这种观点引入心理学领域，并以态度的自动化加工为基础，通过分类任务来考察两种概念（属性维度和目标概念维度）之间自动化联系的紧密程度。在内隐联想测验中，属性维度和目标概念维度分为相容与不相容两种关系。所谓相容，是指属性维度和目标概念维度的关系与个体的内隐态度相一致，如目标概念维度是"女性"，属性维度是"胆小的"；所谓不相容，是指属性维度和目标概念维度之间的关系与个体的内隐态度不一致，甚至相反，如目标概念维度是"女性"，属性维度是"勇敢的"。内隐联想测验的每一部分均有相应的要求，其基本步骤如下：（1）呈现属性维度，被试按要求按键进行分类；（2）呈现目标概念维度，被试按要求按键进行分类；（3）相容联合辨别任务，呈现属性维度和目标概念维度，被试按要求按键进行分类；（4）呈现属性维度，按键要求与（1）相反；（5）相反联合辨别任务，呈现属性维度和目标概念维度，按键要求与（3）相反。后来，Greenwald 对内隐联想测验的基本步骤进行了改进，将其扩展为 7 个步骤，第 4 步骤重复第 3 步骤，第 5 步骤呈现属性维度，第 6 步骤为相反联合辨别任务，第 7 步骤重复第 6 步骤。这样，在改进后的内隐联想测验中，第 1、2、5 步骤可以视为练习部分，第 3、4、6、7 步骤为正式实验部分。[③]

　　Greenwald 等人认为，如果两类概念联系紧密，那么人们在分

①　Schneider D. J. , *The Psychology of Stereotyping*, New York：Guilford, 2004.

②　Greenwald A. G. , McGhee D. E. , & Schwartz J. L. K. , "Measuring individual differences in implicit cognition：The implicit association test", *Journal of Personality and Social Psychology*, Vol. 74, No. 6, October 1998, pp. 1464 – 1480.

③　Greenwald A. G. , McGhee D. E. , & Schwartz J. L. K. , "Measuring individual differences in implicit cognition：The implicit association test", *Journal of Personality and Social Psychology*, Vol. 74, No. 6, March 1998, pp. 1464 – 1480.

类任务中将二者分为一类的速度要快得多。比如，在他们的实验中，目标概念维度为"花、昆虫""乐器、武器"；属性维度为积极词（如"和平、快乐"）与消极词（如"腐烂的、丑陋的"）。结果发现，当被试把"花、乐器"与积极词归为一类，把"昆虫、武器"与消极词归为一类时，被试的分类速度要比相反的分类（把"花、乐器"与消极词归为一类，把"昆虫、武器"与积极词归为一类）快得多。①

　　将这种方法扩展到刻板印象领域可以推测，以性别刻板印象为例，大多数人认为，女性是胆小的，男性是勇敢的，那么将女性与"胆小的"相联系，将男性与"勇敢的"相联系要容易一些，而将女性与"勇敢的"相联系，将男性与"胆小的"相联系要困难一些，表现在反应时上则是前者的反应时要短一些，而后者的反应时相对长一些。

　　Greenwald 等人起初将相容联合辨别任务中的反应时与相反联合辨别任务中的反应时之差作为内隐联想测验的指标，后来又提出新的分析数据的方式。新的分析数据方式如下：（1）只分析第3、4、6、7 步的数据；（2）删除超过 10000ms 的数据；（3）如果一个被试小于 300ms 的反应时占 10% 以上，则剔除这个被试；（4）分别计算第 3 和第 6 两个步骤的总体标准差 ST_1，第 4 和第 7 两个步骤的总体标准差 ST_2；（5）计算每个步骤中正确反应的平均反应时；（6）对于错误的反应，用每个步骤的平均反应时加上 600ms 来代替其反应时；（7）在错误反应的数据被替换后，计算每个步骤的新的平均反应时；（8）分别计算第 6 步与第 3 步的平均反应时之差 M_1，第 7 步与第 4 步的平均反应时之差 M_2；（9）用平均数之差除

---

① Greenwald A. G., McGhee D. E., & Schwartz J. L. K., "Measuring individual differences in implicit cognition: The implicit association test", *Journal of Personality and Social Psychology*, Vol. 74, No. 6, April 1998, pp. 1464 – 1480.

以标准差，即用 M_1 除以 ST_1 得到 D_1，用 M_2 除以 ST_2 得到 D_2；（10）将 D_1 与 D_2 平均，得到 D 值。将 D 值作为最终考察的指标。[①]

为了考察内隐联想测验是否受不同因素的影响，Greenwald 等人对这些可能影响内隐联想测验的因素进行了分析。最终发现，属性维度标签和目标概念维度标签在屏幕上方的位置（左、右），相容辨别任务与相反辨别任务出现的先后顺序，按键的选择，属性维度和目标概念维度出现的时间长短，各阶段刺激词的数量，对错误反应的处理方式等均对结果没有显著的影响。[②] Dasgupta，McGhee 和 Greenwald 等又深入探讨了被试对实验材料（概念词与属性词）的熟悉程度对实验结果的影响，结果表明，被试对实验材料熟悉程度的高低对结果没有显著影响。这些发现均说明，内隐联想测验具有较高的稳定性。[③]

随后，Greenwald 等人又进一步考察了外显测量和内隐测量的结果之间的关系，以揭示内隐联想测验的效度。相关矩阵分析结果显示，各指标之间存在一定程度的相关，说明外显测量和内隐测量都测量到了某一共同的结构，这表明，内隐联想测验具有一定的聚合效度。[④] 同时，不同外显测量和内隐测量内部均具有较

---

① Greenwald A. G. , McGhee D. E. , & Schwartz J. L. K. , "Measuring individual differences in implicit cognition：The implicit association test", *Journal of Personality and Social Psychology*, Vol. 74, No. 6, March 1998, pp. 1464 – 1480.

② Greenwald A. G. , McGhee D. E. , & Schwartz J. L. K. , "Measuring individual differences in implicit cognition：The implicit association test", *Journal of Personality and Social Psychology*, Vol. 74, No. 6, September 1998, pp. 1464 – 1480.

③ Dasgupta N. , McGhee D. E. , Greenwald A. G. , & Banaji M. R. , "Automatic preference for White Americans：Eliminating the familiarity explanation", *Journal of Experimental Social Psychology*, Vol. 36, No. 3, September 2000, pp. 316 – 328.

④ Greenwald A. G. , Nosek B. A. , & Banaji M. R. , "Understanding and using the implicit association test：Ⅰ. An improved scoring algorithm", *Journal of Personality and Social Psychology*, Vol. 85, No. 2, March 2003, pp. 197 – 216.

高的相关，而两测验群之间的相关却相对较低，表明内隐联想测验具有较好的区分效度。[1]

自 1998 年引入 IAT 以来，基于 IAT 的研究如雨后春笋般涌现。相关研究表明，IAT 已经适用于许多不同的偏见（如，性别歧视、异性歧视、年龄歧视等）；儿童研究表明，他们在很小的时候就产生了偏见。[2][3] 因此，至少在一段时间内，内隐联想测验具有较好的信度和效度，是考察内隐社会认知较好的研究方法。

尽管已有研究证明了 IAT 的可靠性和有效性，但还是引起了相当大的争议。一些研究者对 IAT 的"内隐性"提出质疑，即受控制的加工是否会影响 IAT 的表现。事实上，与其他内隐测量法相比，在一些研究中能够观察到 IAT 与显性偏见测量法之间的对应关系[4]。也有研究表明，动机过程[5]和认知控制[6]与 IAT 得分相混淆；一些

① 蔡华俭：《Greenwald 提出的内隐联想测验介绍》，《心理科学进展》2003 年第 3 期。

② Baron A. S. , & Banaji M. R. "The development of implicit attitudes：Evidence of race evaluations from ages 6, 10 & adulthood", *Psychological Science*, No. 17, April 2006, pp. 53 – 58.

③ Newheiser A. , & Olson K. R. , "White and Black American children's implicit intergroup bias", *Journal of Experimental Social Psychology*, No. 48, March 2012, pp. 264 – 270.

④ McConnell A. R. , & Leibold J. M. , "Relations between the Implicit Association Test, explicit racial attitudes, and discriminatory behavior", *Journal of Experimental Social Psychology*, No. 37, October 2001, pp. 435 – 442.

⑤ Vanman E. J. , Saltz J. L. , Nathan L. R. , & Warren J. A. , "Racial discrimination by low-prejudiced Whites：Facial movements as implicit measures of attitudes related to behavior", *Psychological Science*, No. 15, May 2004, pp. 711 – 714.

⑥ Siegel E. F. , Dougherty M. R. , & Huber D. E. , "Manipulating the role of cognitive control while taking the implicit association test", *Journal of Experimental Social Psychology*, No. 48, September 2012, pp. 1057 – 1068.

研究者甚至认为，IAT 得分可以通过简单的指令伪造。[1][2]

其他研究表明，IAT 不能预测相关行为[3]，甚至不能预测其他内隐测量所能预测的行为[4]。元分析表明，IAT 得分与刻板印象领域的行为之间存在可靠的关系，但最近的另一项元分析却发现两者之间的关系很小。[5]

上述研究表明，有关 IAT 的结果有效性还存在争议，也就是说，IAT 有时可能无法可靠地预测行为，其中一个原因在于，它有可能受到"人际之外"联想的影响，这些联想在记忆中存在，但并不影响个人的态度。[6][7]"人际之外"联想可能源于对他人态度、文化规范或其他来源的了解，但它们的存在并不要求在自己

①　De Houwer J. , Beckers T. , & Moors A. , "Novel attitudes can be faked on the Implicit Association Test", *Journal of Experimental Social Psychology*, No. 43, March 2007, pp. 972 – 978.

②　Wallaert M. , Ward A. , & Mann T. , "Explicit control of implicit responses: Simple directives can alter IAT performance", *Social Psychology*, No. 41, September 2010, p. 152.

③　Karpinski A. , & Hilton J. L. , "Attitudes and the Implicit Association Test", *Journal of Personality and Social Psychology*, No. 81, May 2001, pp. 774 – 778.

④　Vanman E. J. , Saltz J. L. , Nathan L. R. , & Warren J. A. , "Racial discrimination by low-prejudiced Whites: Facial movements as implicit measures of attitudes related to behavior", *Psychological Science*, No. 15, November 2004, pp. 711 – 714.

⑤　Oswald F. L. , Mitchell G. , Blanton H. , Jaccard J. , & Tetlock P. E. , "Predicting ethnic and racial discrimination: A meta-analysis of IAT criterion studies", *Journal of Personality and Social Psychology*, No. 105, September 2013, pp. 171 – 192.

⑥　Han H. A. , Olson M. A. , & Fazio R. H. , "The influence of experimentally-created extrapersonal associations on the Implicit Association Test", *Journal of Experimental Social Psychology*, No. 42, April 2006, pp. 259 – 272.

⑦　Olson M. A. , & Fazio R. H. , "Reducing the influence of extra-personal associations on the Implicit Association Test: Personalizing the IAT", *Journal of Personality and Social Psychology*, No. 86, May 2004, pp. 653 – 667.

和自己的文化之间划出一条明显的界限。[1] 例如,Han 等人的研究表明,IAT 评估的态度估计值会受到与态度无关的来源的影响。[2] Han 等人在研究中采用了一种"个性化"的 IAT,其中包括对类别标签的更改(例如,用"我喜欢/我不喜欢"代替"高兴/不高兴"),这样足以减少个人外在联想的影响,并增加与相关行为的对应性。[3] 此后,多项研究表明,个性化 IAT 具有可靠的预测能力。[4][5][6][7] Yoshida 及其同事进一步扩展了自我态度与他人态度之间的区别,引入了一种规范性 IAT,其类别标签包括"大多数人喜欢"和"大多数人不喜欢",试图对社会规范的认知进行隐性评

① Banaji M. R. , Nosek B. A. , & Greenwald A. G. , "No place for nostalgia in science: A response to Arkes and Tetlock", *Psychological Inquiry*, No. 15, Octorber 2004, pp. 279 – 310.

② Han H. A. , Olson M. A. , & Fazio R. H. , "The influence of experimentally-created extrapersonal associations on the Implicit Association Test", *Journal of Experimental Social Psychology*, No. 42, March 2006, pp. 259 – 272.

③ Olson M. A. , & Fazio R. H. , "Reducing the influence of extra-personal associations on the Implicit Association Test: Personalizing the IAT", *Journal of Personality and Social Psychology*, No. 86, September 2004, pp. 653 – 667.

④ Gawronski B. , Deutsch R. , Mbirkou S. , Seibt B. , & Strack F. , "When 'just say no' is not enough: Affirmation versus negation training and the reduction of automatic stereotypes activation", *Journal of Experimental Social Psychology*, No. 44, June 2008, pp. 370 – 377.

⑤ Todd A. R. , Bodenhausen G. V. , Richeson J. A. , & Galinsky A. D. , "Perspective taking combats automatic expressions of racial bias", *Journal of Personality and Social Psychology*, No. 100, July 2011, pp. 1027 – 1042.

⑥ De Houwer J. , Custers R. , & De Clercq A. , "Do smokers have a negative implicit attitude toward smoking?" *Cognition and Emotion*, No. 20, January 2006, pp. 1274 – 1284.

⑦ Houben K. , & Wiers R. W. , "Are drinkers implicitly positive about drinking alcohol? Personalizing the alcohol-IAT to reduce negative extrapersonal contamination", *Alcohol and Alcoholism*, No. 42, September 2007, pp. 301 – 307.

估，也得到具有较高说服力的结果。[1]

（三）反应/不反应任务（The Go / No-go Association Task，GNAT）

除 IAT 之外，内隐社会认知领域的研究者还开发了多种其他改良的 IAT，以解决 IAT 的相对测量问题。例如，Nosek 和 Banaji 研发了 GNAT 任务，这种任务是一种间接评估态度的方法，不需要两个对比类别的态度对象[2]（另见 Williams & Kaufmann，2012）。在标准 IAT 中，无法避免的是需要决定如何对相容和不相容区块类型进行排序或平衡施测，而且经常会观察到排序效应[3]，GNAT 则能够很好地解决这一问题。

GNAT 由内隐联想测验发展而来，是内隐联想测验的一种变式。它本身并不是对内隐联想测验的否定，而是对其的补充和发展。GNAT 涉及考察目标类别与属性概念之间的联结强度，弥补了内隐联想测验中需要提供类别维度，不能对某一对象做出评价的不足。[4] GNAT 结合了信号检测论的内容，实验中包括目标刺激和分心刺激，当出现目标刺激时按空格键反应（称为 Go），当出现分心刺激时不做任何反应（称为 No-go）。

在 Nosek 和 Banaji 的实验中，目标类别为"fruits"和"bugs"，

① Yoshida E., Peach J. M., Zanna M. P., Spencer S. J., "Not all automatic associations are created equal: How implicit normative evaluations are distinct from implicit attitudes and uniquely predict meaningful behavior", *Journal of Experimental Social Psychology*, No. 48, June 2012, pp. 694 – 706.

② Nosek B. A., & Banaji M. R., "The go/no-go association task", *Social Cognition*, No. 19, September 2001, pp. 625 – 666.

③ Nosek B. A., Greenwald A. G., & Banaji M. R., "Understanding and using the Implicit Association Test: II. Method variables and construct validity", *Personality and Social Psychology Bulletin*, No. 31, March 2005, pp. 166 – 180.

④ 梁宁建、吴明证、高旭成：《基于反应时范式的内隐社会认知研究方法》，《心理科学》2003 年第 2 期。

属性类别为"good"和"bad"。在第一阶段中，被试对"fruit"和"good"做出反应，对"bugs"和"bad"不做任何反应；在第二阶段中，被试要对"fruit"和"bad"做出反应，对"bugs"和"good"不做反应。研究采用 d' 为指标，正确的"go"反应为击中，不正确的"go"反应为虚报，然后将击中和需要转化为 Z 分数后，二者的差值即为 d'。d' 代表被试从噪音中区分信号的能力。结果发现，当要求被试对"fruit"和"good"做出反应，其反应时要比对"fruit"和"bad"做出反应慢得多。[1]

（四）其他测量方法

由于 IAT 和启动测量的流行，其他内隐测量可能没有得到应有的关注或使用。[2] 例如，Chen 和 Bargh 开发了一种方法，利用被试接近或回避有价值对象的自动倾向，同时利用运动反应为基础的内隐测量方法。[3] 在这种测量方法中，测量指标可能存在"射手偏差"[4] 和"武器效应"[5]。这两项任务都是启动任务，但

---

① Nosek B. A. , & Banaji M. R. , "The go/no-go association task", *Social Cognition*, No. 19, June 2001, pp. 625 – 666.

② De Houwer, J. , "The extrinsic affective Simon task", *Experimental Psychology*, No. 50, 2003, pp. 77 – 85.

③ Chen M. , & Bargh J. A. , "Consequences of automatic evaluation：Immediate behavioral predispositions to approach or avoid the stimulus", *Personality and Social Psychology Bulletin*, No. 25, July 1999, pp. 215 – 224.

④ Correll J. , Park B. , Judd C. M. , & Wittenbrink B. , "The police officer's dilemma：Using ethnicity to disambiguate potentially threatening individuals", *Journal of Personality and Social Psychology*, No. 83, 2010, p. 1314; Cottrell, C. A. , Richards, D. A. R. , & Nichols, A. L. , "Predicting policy attitudes from general prejudice versus specific intergroup emotions", *Journal of Experimental Social Psychology*, No. 46, February 2002, pp. 247 – 225.

⑤ Payne B. K. , "Prejudice and perception：the role of automatic and controlled processes in misperceiving a weapon", *Journal of Personality and Social Psychology*, No. 81, October 2001, p. 181.

并不是直接测量对偏见目标的评价反应，而是在任务中分别去评估是否"射杀"一个持有武器的人的决定，或者是否将一个工具视为武器的决定。这两者都揭示了对黑人和其他被污名化目标的偏见。

其他内隐测量法不需要配备计算机以记录反应时，因此有可能在实地得到更广泛的应用。[①] 例如，单词片段完成度测量方法已经得到了广大研究者很好的理解和验证。[②③] 其他非基于计算机的内隐测量则倾向于考察被试在描述内群体和外群体成员时使用语言的细微差别。[④⑤]

随着脑科学的兴起以及技术的更新，脑成像和其他生理方法也被用来评估偏见和刻板印象。例如，有研究者采用眨眼惊跳反应来考察黑人与白人面孔的各种动机取向[⑥]；另有研究采用肌电

---

① Vargas P. T., Sekaquaptewa D., & von Hippel W., "Armed only with paper and pencil：'Low-tech' measures of implicit attitudes", in B. Wittenbrink & N. Schwarz (Eds.), *Implicit Measures of Attitudes*, New York：Guilford Press, 2007, pp. 103 – 124.

② Dovidio J. F., Kawakami K., Johnson C., Johnson B., & Howard A., "On the nature of prejudice：Automatic and controlled processes", *Journal of Experimental Social Psychology*, No. 33, May 1997, pp. 510 – 540.

③ Son Hing L. S., Li W., & Zanna M. P., "Inducing hypocrisy to reduce prejudicial responses among aversive racists", *Journal of Experimental Social Psychology*, No. 38, February 2002, pp. 71 – 78.

④ Sekaquaptewa D., Espinoza P., Thompson M., vargas P., & von Hippel W., "Stereotypic explanatory bias：Implicit stereotyping as a predictor of discrimination", *Journal of Experimental Social Psychology*, No. 39, August 2003, pp. 75 – 82.

⑤ von Hippel W., Sekaquaptewa D., & Vargas P., "The linguistic intergroup bias as an implicit indicator of prejudice", *Journal of Experimental Social Psychology*, No. 33, September 1997, pp. 490 – 509.

⑥ Amodio D. M., Harmon-Jones E., & Devine P. G., "Individual differences in the activation and control of affective race bias as assessed by startle eyeblink responses and self-report", *Journal of Personality and Social Psychology*, No. 84, June 2003, pp. 738 – 753.

图（EMG）作为指标，利用被试的细微面部反应考察被试对黑人和白人的情感表现①②③。脑电图（EEG）也常常被用于揭示偏见和刻板印象反应中经常涉及的一些问题④；功能性磁共振成像（fMRI）技术表明，杏仁核与对不同性别面孔的情感反应密切相关。在内隐社会认知领域中，研究者仍旧采用上述设备和行为测量指标相结合的方式，共同考察个体的偏见和刻板印象。

第三节　内隐性别刻板印象及其准确性

关于刻板印象和偏见的准确定义是研究者普遍关注的焦点之一。一般来说，偏见是指对某一群体或该群体成员的负面态度，具有消极的意味。研究者对于刻板印象的定义则更加广泛，一般认为，刻板印象是一种知识结构，是对相关群体的心理描绘。⑤刻板印象代表了我们认为某些社会群体或这些群体中个别成员的特征，尤其是那些能使群体相互区别开来、具有较强代表性的特征。简而言之，刻板印象可以被看作当我们想到某个群体时很快

① Stewart T. L., Amoss R. T., Weiner B. A., Elliott L. A., Parrott D. J., Peacock C. M., & Vanman E. J., "The psychophysiology of social action: Facial electromyographic responses to stigmatized groups predict antidiscrimination action", *Basic and Applied Social Psychology*, No. 35, October 2013, pp. 418 – 425.

② Vanman E. J., Paul B. Y., Ito T. A., & Miller N., "The modern face of prejudice and the structural features that moderate the effect of cooperation on affect", *Journal of Personality and Social Psychology*, No. 73, February 1997, pp. 941 – 959.

③ Vanman E. J., Saltz J. L., Nathan L. R., & Warren J. A., "Racial discrimination by low-prejudiced Whites: Facial movements as implicit measures of attitudes related to behavior", *Psychological Science*, No. 15, November 2004, pp. 711 – 714.

④ Amodio D. M., Harmon-Jones E., Devine P. G., Curtin J. J., Hartley S. L., & Covert A. E., "Neural signals for the detection of unintentional race bias", *Psychological Science*, No. 15, September 2004, pp. 88 – 93.

⑤ Lippman W., *Public Opinion*, New York: Harcourt & Brace, 1922.

就会想到的一些特征。比如，我们想到老年人，就会想到他们是行动缓慢的一群人。

尽管刻板印象可以是积极的，但在谈到刻板印象和偏见时，人们往往会想到其消极的一面，比如认为，老人是思维迟缓的，男性比女性更鲁莽，女性比男性更犹豫等。在有关种族方面的研究中，研究者发现，尽管被试表达了对于黑人的积极刻板印象，但对于旁观者来说，他们感受到的是被试仍然持有消极刻板印象。[1][2] 从这些研究中我们可以看到，有问题的并不是刻板印象本身，而是其准确性。

实际上，要想更好地解决刻板印象或偏见的准确性问题是非常困难的，有研究者曾尝试解决这个问题，但得出的结论并不一致[3][4][5]。在某种程度上，由于人们的感知和现实情况的确存在相关性，因此可能大多数群体信念都有其合理的一面。[6] 也就是说，有问题的是使用刻板印象的过程，比如人们的过度概括，而不是持有

①　Czopp A. M. ，"When is a compliment not a compliment? Evaluating expressions of positive stereotypes"，*Journal of Experimental Social Psychology*，Vol. 44，No. 2，May 2008，pp. 413 – 420.

②　Kay A. C. ，Day M. V. ，Zanna M. P. ，& Nussbaum A. D. ，"The insidious (and ironic) effects of positive stereotypes"，*Journal of Experimental Social Psychology*，Vol. 49，No. 2，October 2013，pp. 287 – 291.

③　Jussim L. ，"Accuracy in social perception：Criticisms，controversies，criteria，components，and cognitive processes"，*Advances in Experimental Social Psychology*，No. 37，February 2005，pp. 1 – 93.

④　Ryan C. ，Park B. ，& Judd C. ，"Assessing stereotype accuracy：Implications for understanding the stereotyping process"，in C. N. Macrae，C. Stangor，& M. Hewstone (Eds.)，*Stereotypes and Stereotyping*，New York：Guilford，1996，pp. 121 – 157.

⑤　Lee Y. T. ，Jussim L. J. ，& McCauley C. R. ，*Stereotype Accuracy：Toward Appreciating Group Differences*，Washington，DC：American Psychological Association，1995.

⑥　Jussim L. ，Cain T. R. ，Crawford J. T. ，Harber K. ，Cohen F. ，& Nelson T. ，"The unbearable accuracy of stereotypes"，*Handbook of prejudice，stereotyping，and discrimination*，June 2009，pp. 199 – 227.

刻板印象本身。[①] 由于刻板印象是对于某个群体而言的，所以无论我们的信念多么准确，它都无法描述群体中的每一个成员，因此，根据类别层面的知识对个人做出判断是完全错误的。

在刻板印象领域，如果采用外显的方式进行研究，很可能出现相反的结论，即被试不愿意承认他们有偏见或对某个群体持有成见，部分原因可能是他们的信念事实上已经发生了变化。因此，越来越多的研究者倾向于采用内隐的方式进行考察，相关方法已在前文中论述过，此处不再赘述。考虑到刻板印象指向不同群体时，结果可能会出现差异，也就是说，与内群体相比，当刻板印象描述的群体是外群体时（被试不属于这个群体），结果会存在不同。这就产生了一个概念性问题，即对外部群体的积极评价是否代表偏见：如果对内群体的偏爱不伴有对外群体的贬损，那么这种偏爱真的是一种问题吗？这个问题引发了研究者的思考，因此在一些实验中，研究者将对外部群体的态度与对内部群体的态度进行比较。[②]

一直以来，社会认知领域的研究者始终认为，群体态度和信念在很大程度上与认知有关。[③][④] 刻板印象是与社会类别相关的特征，代表了一种重要的社会知识形式，基于已有的大量研究，研究者已经了解了许多关于它们在认知层面上是如何形成的以及如何表现出来的。一般来说，刻板印象是作为认知结构存在

---

① Fiske S. T. , "Examining the role of intent: Toward understanding its role in stereotyping and prejudice", in J. S. Uleman & J. A. Bargh (Eds.), *Unintended Thought*, New York: Guilford, 1989, pp. 253 – 286.

② Stangor C. , & Leary S. , "Intergroup beliefs: Investigations from the social side", *Advances in Experimental Social Psychology*, No. 38, May 2006, pp. 243 – 283.

③ Lippman W. , *Public Opinion*, New York: Harcourt & Brace, 1922.

④ Katz D. , & Braly K. W. , "Racial stereotypes of one hundred college students", *Journal of Abnormal and Social Psychology*, No. 28, February 1933, pp. 280 – 290.

的，如图式①、原型②和范例③④。这并不意味着这些信念是僵化的，相反，它们会随着社会环境的变化而变化。⑤⑥⑦

　　除对刻板印象本身进行研究之外，研究者还较为关注刻板印象的内容。Fiske 及其同事对刻板印象的基本成分进行分类，重点关注温暖和能力这两个维度。⑧⑨⑩ 这两个维度是社会心理学有关刻板印象的基本维度⑪，也是在现实世界中真实存在的情况。

① Cox W. T. , Abramson L. Y. , Devine P. G. , & Hollon S. D. , "Stereotypes, prejudice, and depression: The integrated perspective", *Perspectives on Psychological Science*, Vol. 7, No. 5, June 2012, pp. 427 – 449.

② Brewer M. B. , Dull L. , & Lui L. , "Perceptions of the elderly: Stereotypes as prototypes", *Journal of Personality and Social Psychology*, No. 41, July 1981, pp. 656 – 670.

③ Bodenhausen G. V. , Schwarz N. , Bless H. , & Wanke M. , "Effects of atypical exemplars on racial beliefs: Enlightened racism or generalized appraisals?" *Journal of Experimental Social Psychology*, No. 31, September 1995, pp. 48 – 63.

④ Smith E. R. , & Zárate M. A. , "Exemplar-based model of social judgment", *Psychological Review*, Vol. 99, No. 1, December 1992, pp. 3 – 21.

⑤ Smith E. R. , & Zárate M. A. , "Exemplar and prototype use in social categorization", *Social Cognition*, Vol. 8, No. 3, May 1990, pp. 243 – 262.

⑥ Oakes P. J. , Haslam S. A. , & Turner J. C. , *Sterotyping and Social Reality*, Oxford, UK: Blackwell, 1994.

⑦ Smith E. R. , & Zárate M. A. , "Exemplar-based model of social judgment", *Psychological Review*, Vol. 99, No. 1, 1992, pp. 3 – 21.

⑧ Maner J. K. , Kenrick D. T. , Becker D. V. , Robertson T. E. , Hofer B. , Neuberg S. L. , et al. , "Functional projection: How fundamental social motives can bias interpersonal perception", *Journal of Personality & Social Psychology*, Vol. 88, No. 1, May 2005, pp. 63 – 78.

⑨ Cuddy A. J. , Fiske S. T. , & Glick P. , "Warmth and competence as universal dimensions of social perception: The stereotype content model and the BIAS map", *Advances in Experimental Social Psychology*, No. 40, September 2008, pp. 61 – 149.

⑩ Fiske S. T. , Cuddy A. J. C. , Glick P. , & Xu J. , "A model of (often mixed) stereotype content: Competence and warmth respectively follow from perceived status and competition", *Journal of Personality & Social Psychology*, Vol. 82, No. 6, October 2002, pp. 878 – 902.

⑪ Osgood C. E. , Suci G. J. , & Tannenbaum P. H. , *The Measurement of Meaning*, Urbana, IL: University of Illinois Press, 1957.

同时，刻板印象内容的维度划分也更加引发研究者的思考：这些内容背后是什么，刻板印象和偏见的潜在动机是社会认同吗？[1][2][3][4][5]

除认知成分外，我们的态度在很大程度上还基于我们对社会群体的情感反应[6]，它可能会影响我们对任务对象的分类[7]，并且对刻板印象和偏见存在一定影响。

尽管情感可能比认知更重要，但研究者在很大程度上还是把重点放在后者上。[8] 当前刻板印象和偏见领域的研究范式大多来自认知心理学和实验心理学，其原因可能和测量被试情感或情绪存在难度有关。人们在对社会群体成员做出反应并与之互动时，确实会体验到情绪，但他们在自我报告的测量中却很难表达出来。与评估认知的技术相比，当前对被试情绪的测量技术较为贫乏，且准确度有待提高。当前行为实验更多采用自我

① Abrams D. , & Hogg M. A. （Eds. ）, *Social Identity Theory：Constructive and Critical Advances*, New York：Springer-verlag, 1990.

② Deaux K. , Reid A. , Mizrahi K. , & Ethier K. A. , "Parameters of social identity", *Journal of Personality & Social Psychology*, Vol. 68, No. 2, July 1995, pp. 280 – 291.

③ Ellemers J. , Spears R. , & Doosje B. （Eds. ）, *Social Identity：Context, Commitment, Content*, Oxford, UK：Blackwell, 1999.

④ Jackson J. W. , & Smith E. R. , "Conceptualizing social identity：A new framework and evidence for the impact of different dimensions", *Personality & Social Psychology Bulletin*, Vol. 25, No. 1, June 1999, pp. 120 – 135.

⑤ Roccas S. , & Brewer M. , "Social identity complexity", *Personality & Social Psychology Review*, Vol. 6, No. 2, May 2002, pp. 88 – 106.

⑥ Bodenhausen G. V. , Kramer G. P. , & Sasser K. , "Happiness and stereotypic thinking in social judgment", *Journal of Personality & Social Psychology*, Vol. 66, No. 4, July 1994, pp. 621 – 632.

⑦ Dovidio J. F. , Gaertner S. L. , Isen A. M. , & Lowrance R. , "Group representations and intergroup bias：Positive affect, similarity, and group size", *Personality and Social Psychology Bulletin*, Vol. 21, No. 8, February 1995, pp. 856 – 865.

⑧ Sears D. O. , "College sophomores in the laboratory：Influences of a narrow data base on social psychology's view of human nature", *Journal of Personality and Social Psychology*, No. 51, May 1986, pp. 515 – 530.

报告的形式来探讨被试的情绪强度、效价等信息，未来可以借助认知神经科学技术来更精确地定位与情绪相关的脑区并探索相关的大脑活动强度，借此来评定被试情绪。[1]

第四节　内隐性别刻板印象的相关研究

人们常常以性别、年龄、种族、职业等类别对社会群体成员进行区分，内隐刻板印象的研究领域也主要集中在这几种社会类别上。其中，除种族之外，性别是人们主要依赖的类别线索。自 Banaji 和 Greenwald 提出内隐刻板印象的概念以来，内隐性别刻板印象一直都是社会心理学与认知心理学领域着重探讨的问题之一。[2]

一　内隐刻板印象的产生

在较长一段时间之内，社会心理学家比较关注偏见和歧视对少数族裔群体成员直接的社会和健康影响。研究发现，由于被歧视，很大比例的黑人生活贫困，无法获得高薪工作。[3][4] 来自 1999—2000 年的几项调查表明，非裔美国人在美国几乎所有主要死因中的死亡率都很高[5]；与白人相比，少数种族获得医疗保健的机

① Olsson A. , & Phelps E. A. （Eds. ）, *Understanding Social Evaluations：What We can （and cannot）Learn from Neuroimaging*, in B. Wittenbrink & N. Schwartz （Eds. ）, *Implicit Measures of Attitudes*, New York：Guilford, 2007, pp. 159 – 175.

② Banaji M. R. , & Greenwald A. G. , "Implicit stereotype and unconscious prejudice", in Zanna, M. P. , Olson, J. M. ed. , *The Psychology of Prejudice：The Ontario Symposium*, No. 7, May 1994, pp. 55 – 76.

③ Williams D. R. , & Rucker T. D. , "Understanding and addressing racial disparities in health care", *Health Care Financing Review*, Vol. 21, No. 4, October 2000, pp. 75 – 91.

④ Williams D. R. , & Williams-Morris R. , "Racism and mental health：The African American experience", *Ethnicity and Health*, Vol. 5, No. 3 – 4, February 2000, pp. 243 – 269.

⑤ Williams D. R. , "Race, socioeconomic status, and health：The added effect of racism and discrimination", in N. E. Adler & M. Marmot （Eds. ）, *Socioeconomic Status and Health in Industrial Nations：Social, Psychological, and Biological Pathways*, New York：New York Academy of Sciences, Vol. 896, 1999, pp. 173 – 188.

会更少，获得的医疗保健质量更差[1][2]；黑人在许多情况下接受治疗程序的可能性较小，而且往往得不到必要的治疗，延误诊断，或不能控制慢性疾病[3]。

更为重要的是，歧视还会对受歧视者的身心健康产生负面影响。相关研究发现，经常遭受歧视或其他形式的不公平待遇的受鄙视者报告了更多的心理困扰、抑郁以及较低的生活满意度和幸福感。[4][5][6][7][8] 除对身心健康的影响外，感知或误解歧视还可能导致其他各种结果。比如研究发现，歧视对工作招聘和绩效

[1]　Williams D. R. , "Race, socioeconomic status, and health: The added effect of racism and discrimination", in N. E. Adler & M. Marmot (Eds.), *Socioeconomic Status and Health in Industrial Nations: Social, Psychological, and Biological Pathways*, New York: New York Academy of Sciences, Vol. 896, 1999, pp. 173 – 188.

[2]　Williams D. R. , & Rucker T. D. , "Understanding and addressing racial disparities in health care", *Health Care Financing Review*, Vol. 21, No. 4, December 2000, pp. 75 – 91.

[3]　Bach P. B. , Cramer L. D. , Warren J. L. , & Begg C. B. , "Racial differences in the treatment of early-stage lung cancer", *New England Journal of Medicine*, Vol. 341, No. 16, 1999, pp. 1198 – 1205.

[4]　Anderson N. , & Armstead C. , "Toward understanding the association of socioeconomic status and health: A new challenge for the biopsychosocial approach", *Psychosomatic Medicine*, No. 57, September 1995, pp. 213 – 225.

[5]　Corning A. F. , "Self-esteem as a moderator between perceived discrimination and psychological distress among women", *Journal of Counseling Psychology*, Vol. 49, No. 1, June 2002, pp. 117 – 126.

[6]　Glauser A. S. , "Legacies of racism", *Journal of Counseling & Development*, No. 77, July 1999, pp. 62 – 67.

[7]　Kessler R. C. , Mickelson K. D. , & Williams D. R. , "The prevalence, distribution, and mental health correlated of perceived discrimination in the United States", *Journal of Health and Social Behavior*, No. 40, May 1999, pp. 208 – 230.

[8]　Klonoff E. A. , Landrine H. , & Ullman J. B. , "Racial discrimination and psychiatric symptoms among blacks", *Cultural Diversity and Ethnic Minority Psychology*, Vol. 5, No. 4, August 1999, pp. 329 – 339.

评估有很大影响。[1] 少数群体成员在遭受歧视时会感到被排斥[2]，认为自己是歧视受害者的人可能会开始回避或不信任相关社会类别的成员[3]，这种回避可能会导致个体高估针对他们的歧视程度，从而使他们认为偏见是不可避免的。感知者可能会根据他们的成见和偏见行事，这些个体的社会交往功能会受损，其社交质量往往会下降。[4] 因此，偏见和刻板印象会给受害者带来各种压力。

成为歧视的目标并不总是具有消极后果。首先，在一些特定情况下，被鄙视者可能完全不知道自己是受害者，这就能进行有效的自我保护[5]；其次，认为自己是歧视的受害者可以增加对内群体的认同，从而产生相对来说较为积极的结果。[6][7] 有研究者认

① Riach P. A. , & Rich J. , "Fishing for discrimination", *Review of Social Economy*, Vol. 62, No. 4, February 2004, pp. 465 – 486.

② Schmitt M. T. , Branscombe N. R. , Kobrynowicz D. , & Owen S. , "Perceiving discrimination against one's gender group has different implications for well-being in women and men", *Personality and Social Psychology Bulletin*, Vol. 28, No. 2, September 2002, pp. 197 – 210.

③ Terrell F. , Terrell S. L. , & Miller F. , "Level of cultural mistrust as a function of educational and occupational expectations among Black students", *Adolescence*, No. 28, June 1993, pp. 573 – 578.

④ Crocker J. , Voelkl K. , Testa M. , & Major B. , "Social stigma: The affective consequences of attributional ambiguity", *Journal of Personality and Social Psychology*, No. 60, March 1991, pp. 218 – 228.

⑤ Inzlicht M. , McKay L. , & Aronson J. , "Stigma as ego depletion: How being the target of prejudice affects self-control", *Psychological Science*, Vol. 17, No. 3, May 2006, pp. 262 – 269.

⑥ Branscombe N. R. , Schmitt M. T. , & Harvey R. D. , "Perceiving pervasive discrimination among African Americans: Implications for group identification and well-being", *Journal of Personality & Social Psychology*, Vol. 77, No. 1, August 1999, pp. 135 – 149.

⑦ Schmitt M. T. , Spears R. , & Branscombe N. R. , "Constructing a minority group identity out of shared rejection: The case of international students", *European Journal of Social Psychology*, Vol. 33, No. 1, October 2003, pp. 1 – 12.

为，刻板印象是毒害我们许多社会互动的认知"怪物"，它们会让个体产生自我实现的预言，使个体身上的刻板印象显现出来。[1]实际上，刻板印象甚至成为我们日常用语的一部分。

所有受到刻板印象影响的个体都会产生消极后果吗？研究表明，那些更积极、群体认同感更高的个体受刻板印象和偏见的影响较小。[2][3] 在社会认知中，刻板印象属于快速的自发且自动分类[4]，人们将被激活的刻板印象应用到对他人的判断中。当我们疲劳、分心或自尊心受损时[5]，当事情变得棘手时[6]，或者当我们缺乏做更多事情的动力时，我们往往会更多地使用我们的分类，即刻板印象，因为利用我们的刻板印象来衡量另一个人可能会让我们的生活更轻松，不会过多耗费认知资源。[7]

① Bargh J. , "The cognitive monster: The case against the controllability of automatic stereotype effects", in S. Chaiken & Y. Trope (Eds.), *Dual-process Theories in Social Psychology*, New York: Guilford, 1999, pp. 361 – 382.

② Kaiser C. R. , Major B. , & McCoy S. K. , "Expectations about the future and the emotional consequences of perceiving prejudice", *Personality & Social Psychology Bulletin*, Vol. 30, No. 2, May 2004, pp. 173 – 184.

③ Major B. , Kaiser C. R. , & McCoy S. K. , "It's not my fault: When and why attributions to prejudice protect self-esteem", *Personality & Social Psychology Bulletin*, Vol. 29, No. 6, February 2003, pp. 772 – 781.

④ Banaji M. R. , & Hardin C. D. , "Automatic stereotyping", *Psychological Science*, No. 7, August 1996, pp. 136 – 141.

⑤ Govorun O. , & Payne B. K. , "Ego-depletion and prejudice: Separating automatic and controlled components", *Social Cognition*, Vol. 24, No. 2, September 2006, pp. 111 – 136.

⑥ Stangor C. , & Duan C. , "Effects of multiple task demands upon memory for information about social groups", *Journal of Experimental Social Psychology*, No. 27, May 1991, pp. 357 – 378.

⑦ Allport G. W. , *The Nature of Prejudice*, Reading, MA: Addison-Wesley, 1954.

那么，我们会在什么时候使用刻板印象呢？研究表明，我们特别容易对我们不太了解或不关心的人进行分类。简而言之，当分类是我们所掌握的关于某人的全部信息时，或者当我们对更深入地了解某人并不特别感兴趣时，我们可能会完全利用我们的刻板印象。在其他情况下，当我们非常了解某个人时（例如，我们了解我们的好朋友），我们可能会几乎完全忽略他人的群体成员身份，完全在个人层面上对他们做出反应。[1]

我们也更倾向于使用知觉上突出的类别对个体进行分类。因此，我们经常会根据人的性别、种族、年龄和外貌吸引力进行判断，研究表明，这可能是因为当我们看到其他人时，这些特征对我们来说是显而易见的。[2] 此外，当个人处于其他不同类别成员的环境中时，类别也会变得尤为突出，也就是说，当我们判断的对象是单个人或者少数人群时，我们更容易使用刻板印象。[3][4]

那么，人们是在何时开始使用刻板印象的呢？

研究者对刻板印象的来源做了深入的考证，发现其和认知

① Madon S., Jussim L., Keiper S., Eccles J., Smith A., & Palumbo P., "The accuracy and power of sex, social class, and ethnic stereotypes: A naturalistic study in person perception", *Personality & Social Psychology Bulletin*, Vol. 24, No. 12, June 1998, pp. 1304 – 1318.

② Brewer M. B., "A dual process model of impression formation", in T. K. Srull & R. S. Wyer (Eds.), *Advances in Social Cognition*, Hillsdale, NJ: Erlbaum, Vol. 1, 1988, pp. 1 – 36.

③ Cota A. A., & Dion K. L., "Salience of gender and sex composition of ad hoc groups: An experimental test of distinctiveness theory", *Journal of Personality and Social Psychology*, Vol. 50, No. 4, December 1986, pp. 770 – 776.

④ Oakes P. J., Turner J. C., & Haslam S. A., "Perceiving people as group members: The role of fit in the salience of social categorizations", *British Journal of Social Psychology*, No. 30, May 1991, pp. 125 – 144.

表征的发展有较大的关联。①②③④⑤⑥ 比如，儿童对学习社会分类
和刻板印象，以及了解如何将自己融入这一分类系统有着积极
的、似乎是与生俱来的兴趣。⑦⑧ 因此，儿童很早就学会了刻板
印象，并对其充满信心，认为它不可能改变。而随着认知的发
展，到 10 岁左右儿童会变得更加灵活。⑨⑩

　　有关刻板印象内容的来源，可能和父母、同伴、媒体等社

① Bigler R. , "The role of classification skill in moderating environmental influences on children's gender stereotyping: A study of the functional use of gender in the classroom", *Child Development*, No. 66, March 1995, pp. 1072 – 1087.

② Bigler R. , & Liben L. , "Cognitive mechanisms in children's gender stereotyping: Theoretical and educational implications of a cognitive-based intervention", *Child Development*, No. 63, June 1992, pp. 1351 – 1363.

③ Dunham Y. , & Degner J. , "Origins of intergroup bias: Developmental and social cognitive research on intergroup attitudes", *European Journal of Social Psychology*, Vol. 40, No. 4, July 2010, pp. 563 – 568.

④ Newheiser A. , & Olson K. R. , "White and Black American children's implicit intergroup bias", *Journal of Experimental Social Psychology*, No. 48, May 2012, pp. 264 – 270.

⑤ Pauker K. , Ambady N. , & Apfelbaum E. P. , "Race salience and essentialist thinking in racial stereotype development", *Child Development*, Vol. 81, No. 6, September 2010, pp. 1799 – 1813.

⑥ Ziv T. , & Banaji M. R. , "Representations of social groups in the early years of life", in S. T. Fiske and C. N. Macrae, *The SAGE Handbook of Social Cognition*, London: Sage, 2012, p. 372.

⑦ Ruble D. , & Martin C. , "Gender development", in W. Damon (Ed.), *Handbook of Child Psychology* (5th ed.), New York: Wiley, 1998, pp. 933 – 1016.

⑧ Stangor C. , & Ruble D. N. , "Differential influences of gender schemata and gender constancy on children's information processing and behavior", *Social Cognition*, No. 7, March 1998, pp. 353 – 372.

⑨ Bigler R. , & Liben L. , "Cognitive mechanisms in children's gender stereotyping: Theoretical and educational implications of a cognitive-based intervention", *Child Development*, No. 63, October 1992, pp. 1351 – 1363.

⑩ Signorella M. , Bigler R. , & Liben L. , "Developmental differences in children's gender schemata about others: A meta-analytic review", *Developmental Review*, No. 13, December 1993, pp. 147 – 183.

会环境的影响有关。①②③ 但有关这些社会因素影响的研究并未得到定论，因此，未来研究（比如，双生子研究）还需要继续对社会环境的因素进行更加细致的探讨，来揭示刻板印象的形成在多大程度上和先天或者后天因素有关。④⑤

　　之前的研究表明，刻板印象或者偏见在一定程度上可能和进化有关，比如有研究者认为，我们喜欢那些在我们看来相似的人，因为他们更有可能是乐于助人和善良的人，而鄙视和回避那些可能存在不良社交经历、可能有病或威胁到重要群体价值观的人。⑥⑦⑧⑨在科技高度发展的今天，我们也要承认媒体的重要影响。电影、电

①　Aboud F. E., *Children and Prejudice*, New York: Basil Blackwell, 1988.

②　Aboud F. E., & Amato M., "Developmental and socialization influences on intergroup bias", in R. Brown & S. Gaertner (Eds.), *Blackwell Handbook in Social Psychology*: Vol. 4: Intergroup Processes, New York: Blackwell, 2001, pp. 65 – 85.

③　Stangor C., & Leary S., "Intergroup beliefs: Investigations from the social side", *Advances in Experimental Social Psychology*, No. 38, May 2006, pp. 243 – 283.

④　Olson J. M., Vernon P. A., Harris J. A., & Jang K. L., "The heritability of attitudes: A study of twins", *Journal of Personality and Social Psychology*, Vol. 80, No. 6, March 2001, pp. 845 – 860.

⑤　Lewis G. J., Kandler C., & Riemann R., "Distinct heritable influences underpin in-group love and out-group derogation", *Social Psychological and Personality Science*, Vol. 5, No. 4, August 2014, pp. 407 – 413.

⑥　Collins E. C., Crandall C. S., & Biernat M., "Stereotypes and implicit social comparison: Shifts in comparison-group focus", *Journal of Experimental Social Psychology*, Vol. 42, No. 4, February 2006, pp. 452 – 459.

⑦　Maner J. K., Kenrick D. T., Becker D. V., Robertson T. E., Hofer B., Neuberg S. L., et al., "Functional projection: How fundamental social motives can bias interpersonal perception", *Journal of Personality & Social Psychology*, Vol. 88, No. 1, August 2005, pp. 63 – 78.

⑧　Neuberg S. L., Kenrick D. T., & Schaller M., "Human threat management systems: Self-protection and disease avoidance", *Neuroscience & Biobehavioral Reviews*, Vol. 35, No. 4, May 2011, pp. 1042 – 1051.

⑨　van de Vliert E., "Climato-economic origins of variation in ingroup favoritism", *Journal of Cross-Cultural Psychology*, Vol. 42, No. 3, May 2011, pp. 494 – 515.

视和网络不仅创造了相关的刻板印象,更重要的是它们为我们提供了相关的社会规范,潜在地告诉我们可以喜欢谁,不可以喜欢谁。所以,未来研究可以考察媒体对刻板印象或者群体观念的纵向影响。

二 内隐性别刻板印象的加工特点及机制

内隐性别刻板印象的研究最初来源于 Jacoby,Kelley,Brown 和 Jasechko 关于内隐记忆的实验。[①] 在他们的研究中,首先向被试提供包括一些名人和普通人姓名的名单,被试要仔细阅读这些人的姓名。一天之后,向被试呈现一份名单,这份名单中不仅包括原有的普通人和名人的姓名,也有新加入的普通人和名人的姓名。被试的任务是判断这些人是不是名人。Jacoby 等人认为,由于先前对一些姓名有所熟悉,因此被试可能会把旧的普通人姓名判断为名人。实验结果显示:被试对旧的普通人名字的记忆具有更高的错误率。Banaji 和 Greenwald 对 Jacoby 等人的研究进行改进,将名字的性别作为变量加入分析,结果发现:名字的性别对被试的判断具有重要的影响。具体来说,被试倾向于将男性判断为名人,或者说被试认为,与女性相比,男性更具声望。这个发现为内隐性别刻板印象的研究提供了有力的证据。[②]

随后,大量的研究者投入内隐性别刻板印象的研究中,并取得丰硕成果。比如,Banaji 在其实验中向被试呈现男性、女性、中性

---

① Jacoby L. L. , Kelley C. M. , Brown J. , & Jasechko J. , "Becoming famous overnight. Limits on the ability to avoid unconscious influences of the past", *Journal of personality and Social Psychology*, Vol. 16, No. 2, March 1989, pp. 326 – 338.

② Banaji M. R. , & Greenwald A. G. , "Implicit stereotype and unconscious prejudice", in Zanna, M. P. , Olson, J. M. ed. , "The psychology of prejudice", *The Ontario Symposium*, No. 7, February 1994, pp. 55 – 76.

靶子行为，接着让被试判断靶子的依赖性与攻击性。① 结果发现，与其他两组被试相比，呈现女性靶子行为的被试倾向于将靶子评定为具有依赖性，而呈现男性靶子行为的被试与其他两组被试相比，更倾向于将靶子评定为具有攻击性。

随着相关研究的不断发展，内隐刻板印象的可控性逐渐成为心理学工作者开始关注的焦点。② 传统观点认为，当向个体呈现类别化信息时，与社会类别（social categorization）相关的知识（如，刻板印象）便会被自动激活。由于这种激活是一个自动化过程，因此，它基本不受外在因素的影响，也不会因为时间和环境的变化而改变。③ Devine 和 Dovidio 等认为，尽管人们尝试忽略刻板印象，但由于它的自动化激活过程，人们无法预期它的发生④⑤；Bargh 也认为，即便个体不想产生刻板印象，但是，当其面对类别化信息时，刻板印象都会不由自主地产生⑥。因此，有研究者认为，刻板印象

---

① Banaji M. R. , "Implicit stereotyping in person judgment", *Journal of Exerimental Social Psychology*, Vol. 65, No. 2, October 1993, pp. 272 – 281.

② Bargh J. , "The cognitive monster: The case against the controllability of automatic stereotype effects", in S. Chaiken & Y. Trope (Eds.), *Dual-process Theories in Social Psychology*, New York: Guilford, 1999, pp. 361 – 382.

③ Bargh J. , "The cognitive monster: The case against the controllability of automatic stereotype effects", in S. Chaiken & Y. Trope (Eds.), *Dual-process Theories in Social Psychology*, New York: Guilford, 1999, pp. 361 – 382.

④ Devine P. G. , "Stereotypes and prejudice: Their automatic and controlled components", *Journal of Personality and Social Psychology*, No. 56, May 1989, pp. 5 – 18.

⑤ Dovidio J. , Evans N. , & Tyler R. , "Racail stereotypes: The contents of their cognitive representation", *Journal of Experimental Social Psychology*, No. 22, March 1986, pp. 22 – 37.

⑥ Bargh J. A. , Chen M. , & Burrows L. , "Automaticity of social behavior: Direct effects of trait construct and stereotype activation on action", *Journal of Personality & Social Psychology*, Vol. 71, No. 2, August 1996, pp. 230 – 244.

是难以消除的"认知魔鬼"（*cognitive monster*）。[①]

但是，另一批研究者向传统观点提出了挑战，他们提出一种修正的观点，这种观点认为，尽管刻板印象的激活属于自动化过程，但是这种自动化过程是有条件的，比如，Gilbert 和 Hixon 在其研究中发现，刻板激活需要个体某些注意资源的参与。[②] 随后，根据 Bargh 提出的"条件自动化"的观点，刻板印象不再是"认知魔鬼"，而是可以有效地减弱甚至消除。[③] 或者说，通过控制某些能够影响内隐刻板印象的因素，以达到减弱甚至消除内隐刻板印象的效果。一些研究者认为，当人们被某种动机驱使[④]，或者努力练习某种策略以期避免自动化的偏见时，这种自动化的偏见是可以改变的[⑤][⑥]。然而，有研究者对这些改变自动化偏见的方法进行了比较，他们发现，通过改变个体生存的社会环境，而非直接操纵个体的目

①　Bargh J. , "The cognitive monster: The case against the controllability of automatic stereotype effects", in S. Chaiken & Y. Trope（Eds. ）, *Dual-process Theories in Social Psychology*, New York: Guilford, 1999, pp. 361 – 382.

②　Gilbert D. T. , & Hixon J. G. , "The trouble of thinking: Activation and application of stereotypic beliefs", *Journal of Personality and Social Psychology*, No. 60, June 1991, pp. 509 – 517.

③　Bargh J. , "The cognitive monster: The case against the controllability of automatic stereotype effects", in S. Chaiken & Y. Trope（Eds. ）, *Dual-process Theories in Social Psychology*, New York: Guilford, 1999, pp. 361 – 382.

④　Lowery B. S. , Hardin C. D. , & Sinclair S. , "Social influence effects on automatic racial prejudice", *Journal of Personality and Social Psychology*, No. 81, July 2001, pp. 842 – 855.

⑤　Blair I. V. , Ma J. E. , & Lenton A. P. , "Imaging stereotypes away: The moderation of implicit stereotypes through mental imagery", *Journal of Personality and Social Psychology*, No. 81, September 2001, pp. 828 – 841.

⑥　Kawakami K. , Dovidio J. F. , Moll J. , Hermsen S. , & Russin A. , "Just say no（to stereotyping）: effects of training in the negation of stereotypic associations on stereotypic activation", *Journal of Personality and Social Psychology*, No. 78, May 2000, pp. 871 – 888.

标和动机，自动化偏见是可以被削弱甚至消除的。[1][2]　比如，在Das-gupta 和 Greenwald 的研究中，主试"无意地"向实验组的被试呈现令人尊敬的黑人样例和令人厌恶的白人样例，而向控制组的被试呈现的样例无种族差异，结果发现，与控制组被试相比，实验组的被试表现出较少的种族偏见。更为重要的是，这种种族偏见的改变持续超过 24 小时，并且，这种策略不仅能被应用于种族偏见，还能应用于其他被偏见化的群体，如老年人、女性等。[3]

从上述研究中可以看出，首先，刻板印象并非不可控制的"认知魔鬼"，而是可以通过一定的策略加以改变；其次，研究者通过比较发现，反刻板样例的呈现能够较好地削弱刻板印象。或者说，反刻板印象是控制刻板印象的有效方法。

第五节　性别刻板印象威胁效应及相关研究

一　刻板印象威胁效应概述

刻板印象威胁可以被看作一种情境困境，在这种困境中，个人的行为表现有可能证实他人对其群体的负面刻板印象。[4]　例

---

① Dasgupta N., McGhee D. E., Greenwald A. G., & Banaji M. R., "Automatic preference for White Americans: Eliminating the familiarity explanation", *Journal of Experimental Social Psychology*, Vol. 36, No. 3, August 2000, pp. 316 – 328.

② Macrae C. N., Bodenhausen G. V., & Miline A. B., "The dissection of selection in person perception: Inhibitory processes in social stereotyping", *Journal of Personality and Social Psychology*, Vol. 69, No. 3, June 1995, pp. 397 – 407.

③ Dasgupta N., & Greenwald A. G., "On the malleability of automatic attitudes: Combating automatic prejudice with images of admired and disliked individuals", *Journal of Personality and Social Psychology*, No. 81, February 2001, pp. 800 – 814.

④ Steele C. M., & Aronson J., "Stereotype threat and the intellectual performance of African Americans", *Journal of Personality and Social Psychology*, No. 69, March 1995, pp. 797 – 811.

如，在一段时间内人们普遍认为，相较于男性而言，女性并不擅长数学，因此，当告知女性要进行数学测验，同时她们的成绩要和男性进行比较时，她们的数学成绩往往要比她们真实的数学能力表现得差一些，因为她们可能会害怕证实这种负面的刻板印象。研究者发现，这种对印证刻板印象的恐惧会使最佳表现所需的认知系统"崩溃"，最终导致考试成绩低下。[①②] Steele 和 Aronson 首次在研究中发现了存在于种族刻板印象中的威胁效应。他们的研究表明，如果情境本身造成或放大了成绩的群体差异，那么当情境以不那么刻板的方式呈现时，黑人大学生的成绩应该会比实际成绩好得多。[③] 事实上，在他们的研究中，当这些问题被描述为简单的实验室任务时而非涉及种族刻板印象的任务时，非裔美国人的表现要比当任务被描述为智力诊断测量时好很多。

过去几十年的研究在多个领域印证了刻板印象威胁效应的存在，比如，研究发现，刻板印象威胁效应不仅导致非裔美国人成绩低下[④]，也导致拉美裔美国人[⑤]和社会经济地位低下的人

---

① Schmader T. , Johns M. , & Forbes C. , "An integrated process model of stereotype threat effects on performance", *Psychological Review*, No. 115, June 2008, pp. 336 – 356.

② Schmader T. , & Beilock S. , "An integration of processes that underlie stereotype threat", in M. Inzlicht & T. Schmader (Eds.), *Stereotype Threat：Theory, Process, and Application*, New York：Oxford University Press, 2011.

③ Steele C. M. , & Aronson J. , "Stereotype threat and the intellectual performance of African Americans", *Journal of Personality and Social Psychology*, No. 69, March 1995, pp. 797 – 811.

④ Steele C. M. , & Aronson J. , "Stereotype threat and the intellectual performance of African Americans", *Journal of Personality and Social Psychology*, No. 69, May 1995, pp. 797 – 811.

⑤ Gonzales P. M. , Blanton H. , & Williams K. J. , "The effect of stereotype threat and double-minority status on the test performance of Latino women", *Personality and Social Psychology Bulletin*, No. 28, July 2002, pp. 659 – 670.

群在标准化测试中成绩较低[①]。同时，在数学和科学领域的女性[②]、记忆领域的老年人[③]以及田径领域的白人[④]群体中也发现了刻板印象威胁效应。因此，这是一种在不同群体、不同任务和不同国家中广泛存在的现象。但是一项研究发现，即使是传统意义上不被社会边缘化的群体（比如，白人男性），当他们得知自己的数学成绩将和亚洲人进行比较时，也会表现出这种效应。[⑤]

多数研究表明，刻板印象威胁效应在心理学学术界和普通大众中都非常流行。1995 年，Steele 和 Joshua Aronson 在《人格与社会心理学杂志》（*Journal of Personality and Social Psychology*）上发表了第一篇关于刻板印象威胁的实证文章，如今这篇文章已被广泛引用并被奉为经典之作。自这篇论文发表以来，刻板印象威胁已成为过去几十年中社会心理学领域探讨最热烈的话题之一。但是，对刻板印象威胁的兴趣并不局限于社会心理学家之间的学术讨论。这项研究的受众非常广泛，不仅包括心理学学术界，还包括其他学科，比如社会学、教育学，甚至包括学术界以外的公众。

---

① Croizet J. C. , & Millet M. , "Social class and test performance: From stereotype threat to symbolic violence and vice versa", in M. Inzlicht & T. Schmader (Eds.), *Stereotype Threat: Theory, Process, and Application*, New York: Oxford University Press, 2011.

② Logel C. , Peach J. , & Spencer S. J. , "Threatening gender and race: Different manifestations of stereotype threat", in M. Inzlicht & T. Schmader (Eds.), *Stereotype Threat: Theory, Process, and Application*, New York: Oxford University Press, 2011.

③ Chasteen A. L. , Kang S. K. , & Remedios J. D. , "Aging and stereotype threat: Development, process, and interventions", in M. Inzlicht & T. Schmader (Eds.), *Stereotype Threat: Theory, Process, and Application*, New York: Oxford University Press, 2011.

④ Stone J. , Chalabaev A. , & Harrison C. K. , "The impact of stereotype threat on performance in sports", in M. Inzlicht & T. Schmader (Eds.), *Stereotype Threat: Theory, Process, and Application*, New York: Oxford University Press, 2011.

⑤ Aronson J. , Lustina M. J. , Good C. , Keough K. , Steele C. M. , & Brown J. , "When white men can't do math: Necessary and sufficient factors in stereotype threat", *Journal of Experimental Social Psychology*, No. 35, May 1999, pp. 29 – 46.

刻板印象威胁效应也具有广泛的适用性。自从第一篇有关刻板印象威胁效应的论文发表后，越来越多的研究者在不同领域发现了此效应。比如，研究者在学校中的黑人和女性、女性汽车司机[1]和白人运动员[2]等群体中均考察了这个效应及其相关的心理机制和生物机制。[3][4]

二 刻板印象威胁效应的类别

刻板印象威胁效应是指人们担心自己的表现或行为会被负面刻板印象透视[5][6][7][8]，这种担心会扰乱和影响人们在负面刻板印象领

[1] Yeung N. C. J. , & von Hippel C. , "Stereotype threat increases the likelihood that female drivers in a simulator run over jaywalkers", *Accident Analysis & Prevention*, No. 40, September 2008, pp. 667 – 674.

[2] Stone J. , Chalabaev A. , & Harrison C. K. , "The impact of stereotype threat on performance in sports", in M. Inzlicht & T. Schmader (Eds.), *Stereotype Threat: Theory, Process, and Application*, New York: Oxford University Press, 2011.

[3] Schmader T. , & Beilock S. , "An integration of processes that underlie stereotype threat", in M. Inzlicht & T. Schmader (Eds.), *Stereotype Threat: Theory, Process, and Application*, New York: Oxford University Press, 2011.

[4] Mendes W. B. , & Jamieson, "Embodied stereotype threat: Exploring brain and body mechanisms underlying performance impairments", in M. Inzlicht & T. Schmader (Eds.), *Stereotype Threat: Theory, Process, and Application*, New York: Oxford University Press, 2011.

[5] Aronson J. , Lustina M. J. , Good C. , Keough K. , Steele C. M. , & Brown J. , "When white men can't do math: Necessary and sufficient factors in stereotype threat", *Journal of Experimental Social Psychology*, No. 35, March 1999, pp. 29 – 46.

[6] Shapiro J. R. , & Neuberg S. L. , "From stereotype threat to stereotype threats: Implications of a multi-threat framework for causes, moderators, mediators, consequences, and interventions", *Personality and Social Psychology Review*, No. 11, August 2007, pp. 107 – 130.

[7] Steele C. M. , "A threat in the air: How stereotypes shape intellectual identity and performance", *American Psychologist*, No. 52, May 1997, pp. 613 – 629.

[8] Steele C. M. , Spencer S. J. , & Aronson J. , "Contending with group image: The psychology of stereotype and social identity threat", in M. P. Zanna (Ed.), *Advances in Experimental Social Psychology*, . San Diego: Academic Press, Vol. 34, 2002, pp. 379 – 440.

域的表现[①]。例如，在受到刻板印象威胁的情况下，女性在数学和科学领域的表现比其真实成绩更低一些[②]，少数族裔/种族学生在学术任务中表现不佳[③]，男性在社会敏感度测量中表现不佳[④]。在过去的几十年里，刻板印象威胁研究呈爆炸式增长，数千项研究对不同的刻板印象、不同的群体和不同的边界条件进行了调查，揭示了这种现象的深远影响。然而，关于刻板印象威胁效应的心理基础是什么却并未得到很好的回答。有研究者提出疑问：如果刻板印象威胁是一种对被视为刻板印象的担忧，那么一个人可能会在谁的心目中担心这些刻板印象被证实呢？举例来说，如果有关女性数学的刻板印象威胁效应是指女性害怕自己被带上"数学不好的帽子"，那么女性在考虑到谁的时候害怕被带上这个帽子呢？是女性群体，是男性群体，还是广泛的社会大众？抑或是三者都有可能？

Shapiro 和 Neuberg 的多重威胁框架概述了六种核心的刻板印象威胁效应[⑤]，这些威胁产生于对两个维度的考虑：刻板印象威胁的目标（这一行为反映了谁：自我或自己的群体）和刻板印象威胁的来源（谁会就这一行为得出结论：自我、外群体他人或内群体他

①　Schmader T. , & Beilock S. , "An integration of processes that underlie stereotype threat", in M. Inzlicht & T. Schmader（Eds. ）, *Stereotype Threat: Theory, Process, and Application*, New York: Oxford University Press, 2011.

②　Spencer S. J. , Steele C. M. , & Quinn D. M. , "Stereotype threat and women's math performance", *Journal of Experimental Social Psychology*, No. 35, August 1999, pp. 4 – 28.

③　Steele C. M. , & Aronson J. , "Stereotype threat and the intellectual performance of African Americans", *Journal of Personality and Social Psychology*, No. 69, June 1995, pp. 797 – 811.

④　Koenig A. M. , & Eagly A. H. , "Stereotype threat in men on a test of social sensitivity", *Sex Roles*, No. 52, March 2005, pp. 489 – 496.

⑤　Shapiro J. R. , & Neuberg S. L. , "From stereotype threat to stereotype threats: Implications of a multi-threat framework for causes, moderators, mediators, consequences, and interventions", *Personality and Social Psychology Review*, No. 11, July 2007, pp. 107 – 130.

人）。这些维度相互交叉产生了不同类型的刻板印象威胁效应。

（一）自我概念威胁

自我概念威胁是指对"自己眼中"的刻板化特征的恐惧，也就是害怕看到自己实际上拥有负面的刻板化特征。例如，一名女性内心可能存在这样的潜在假设：由于性别原因，我不如男性数学成绩好。由于对数学成绩不佳的担心支持了潜伏在她内心深处的假设，因此，她的自我概念威胁风险会增加到她认同负面刻板印象的领域（例如，将数学视为她身份的核心和重要部分），并相信负面刻板印象是真的。也就是说，如果这位女性认为刻板印象没有可信度，那么她就没有理由害怕拥有这种被认为"数学不如男性"的负面特征。

（二）群体概念威胁

群体概念威胁是指害怕看到自己所属的群体拥有负面的刻板印象特质，也就是说，个体害怕自己的表现会在自己心中证实这个群体被合理性地贬低。举例来说，一位女性可能会担心自己在数学考试中的不佳表现会在她自己心中证实女性的数学能力不如男性。那么，如果一个人认同自己的群体，将自己视为群体的代表并且关心自己关于群体的看法对个人自我成长的影响，那么他受到群体概念威胁的风险就会增加。此外，与自我概念威胁类似，如果一个人相信刻板印象可能是真的，那么受到群体概念威胁的风险就会增加。

（三）自我声誉威胁

自我声誉威胁是指对他人眼中的刻板印象的恐惧，也就是说，害怕他人因为对自己负面的刻板印象而受到其评判或不好的待遇。例如，高中女生可能会担心因数学考试成绩不佳而导致父母、老师根据有关女性数学能力的刻板印象来判断她，最终导致在高考选择专业时可能会阻止她选择与数学有关的专业。因此，与自我概念威胁相比，自我声誉威胁是一种更公开的刻板印象威胁，当一个人认为观众可以识别他是否属于刻板印象群体时，就会出现这种威胁。所以，这位高中女生的自我声誉威胁会让她认为他人存在"女性数学能力不如男性"的刻板印象，如果她的数学成绩不公开，或者别人并不知道她的数学成绩，那么她的自我声誉威胁将会降低。也就是说，只要这位女同学关心她的行为对他人如何看待她的影响，她就会面临自我声誉威胁的风险。

（四）群体声誉威胁

群体声誉威胁是指害怕在他人心目中强化对自己群体的负面刻板印象，也就是个体害怕成为自己群体的"不良典范"。上述例子中的高中女生，此时害怕数学考试成绩不佳会在别人心中强化对女性整体及其数学能力的负面刻板印象。因此，一个人在刻板印象领域的表现必须让他人看到，并与刻板印象群体有明显的联系，才会出现群体声誉威胁。此外，如果一个人认为在他人心目中刻板印象是真实的，或可能是真实的，那么他遭受群体声誉威胁的风险就会增加。例如，如果这位女生参加了数学考试，但没有提供任何个人信息，那么她就不应该担心自己在群体中的表现不佳，因为别人并不知道她的成绩，也无法将她的成绩与女性整个群体联系起来。此外，对负面刻板印象群体的认同会增加一个人受到群体声誉威胁的风险，如果一个人不把群体视为中心身份，他就不太可能在意让这个群体失望。

综上所述，鉴于不同的条件组合会引发每一种刻板印象威胁，因此，每一种刻板印象威胁都可以独立于其他威胁而存在。举例来说，如果一个女性认同数学领域，但怀疑关于女性和数学的负面刻板印象是否真的存在，那么她就有可能受到自我概念威胁，但如果她不在意别人如何看待、评价或对待她，并且从不公开参与与数学相关的任务，那么她就不会受到其他刻板印象威胁。相反，如果一个女性认为男性对女性持有与数学相关的负面刻板印象，并认为她的数学成绩会与她个人有关，那么如果她非常不关心自己的群体（群体认同度低）和数学（领域认同度低），她就有可能受到自我归因威胁（外群体），但不会受到其他刻板印象威胁。

三 刻板印象威胁效应的影响因素

导致刻板印象威胁效应的不同情境因素（比如，个人表现的私密性与公开性）和性格因素（比如，群体认同）表明，不同的变量会促进或抑制对不同刻板印象威胁的体验。通过前文对刻板印象威胁效应的类别描述可知，对刻板化领域的认同、相信刻板印象、认同自身所在的群体、认同刻板化行为与自身有关等，都可能会诱发刻板印象威胁效应。除这些因素之外，与每种刻板印象威胁具体相关的其他因素也可能会导致刻板印象威胁的产生。例如，以自我为目标的刻板印象威胁（自我概念威胁、自身声誉威胁）可能在一定程度上受到与自己或他人给予正面评价的愿望相关的变量的影响，包括自尊水平、对自身声誉的需求、对他人的期望等①②③，而群体即目标的刻板印象威胁（群体概念威胁、

---

① Crocker J., & Wolfe C. T., "Contingencies of self-worth", *Psychological Review*, No. 108, 2001, pp. 593 – 623.

② Kernis M. H., Paradise A. W., Whitaker D. J., Wheatman S. R., & Goldman B. N., "Master of one's psychological domain? Not likely if one's self-esteem is unstable", *Personality and Social Psychology Bulletin*, No. 26, August 2000, pp. 1297 – 1305.

③ Leary M. R., Kelly K. M., Cottrell C. A., & Schreindorfer L. S., *Individual Differences in the Need to Belong: Mapping the Nomological Network*, unpublished manuscript, Wake Forest University, 2006.

群体声誉威胁）则在一定程度上受到更普遍地与希望自己的群体被正面看待的变量影响，比如，集体主义情境的激活、个人隐私威胁、群体名誉威胁等。[1][2] 还有研究者认为，自我监控、公众自我评价等也会导致刻板印象威胁的产生。[3][4][5][6]

上述刻板印象威胁都会在大多数负面刻板印象领域产生类似的行为结果，导致个体在相关领域表现不佳，但每一种刻板印象威胁与这一结果之间关系的机制可能存在区别。研究表明，由刻板印象威胁引起的表现下降可能存在一个复杂的中介过程：早期的中介因素，如消极的侵入性想法和消极的情绪反应，会降低工作记忆的效率，这是导致个体表现下降的最重要原因。[7][8] 虽然工

---

① Luhtanen R. , & Crocker J. , "A collective self-esteem scale: Self-evaluation of one's social identity", *Personality and Social Psychology Bulletin*, No. 18, 1992, pp. 302 –318.

② Sellers R. M. , Rowley S. A. J. , Chavous T. M. , Shelton J. N. , & Smith M. , "Multidimensional inventory of black identity: Preliminary investigation of reliability and construct validity", *Journal of Personality and Social Psychology*, No. 73, June 1997, pp. 805 – 815.

③ Briggs S. R. , Cheek J. M. , & Buss A. H. , "Other directedness question-naire", *Journal of Personality and Social Psychology*, No. 38, May 1980, pp. 679 – 686.

④ Inzlicht M. , Aronson J. , Good C. , & McKay L. , "A particular resiliency to threatening environments", *Journal of Experimental Social Psychology*, No. 42, September 2006, pp. 323 – 336.

⑤ Scheier M. F. , & Carver C. S. , "The Self-consciousness Scale: A revised version for use with general populations", *Journal of Applied Social Psychology*, No. 15, October 1985, pp. 687 – 699.

⑥ Briggs S. R. , Cheek J. M. , & Buss A. H. , "Other directedness question-naire", *Journal of Personality and Social Psychology*, No. 38, May 1980, pp. 679 – 686.

⑦ Schmader T. , & Beilock S. , "An integration of processes that underlie stereotype threat", in M. Inzlicht & T. Schmader (Eds.), *Stereotype Threat: Theory, Process, and Application*, New York: Oxford University Press, 2011.

⑧ Schmader T. , Johns M. , & Forbes C. , "An integrated process model of stereotype threat effects on performance", *Psychological Review*, No. 115, May 2008, pp. 336 – 356.

作记忆的减少同样会影响所有刻板印象威胁的表现,但不同刻板印象威胁的早期中介因素有所不同。例如,在以群体为目标的刻板印象威胁中,基于群体的侵入性想法可能会出现,并且会消耗工作记忆;而在以自我为目标的刻板印象威胁中,基于自我的侵入性想法可能会出现,并且这些想法会消耗工作记忆。

除了降低工作记忆的效率,个体可能还会主动调节侵入性想法,因而导致自我控制产生变化。研究表明,为了调节和管理这些干扰性想法和负面情绪而进行的自我控制,会耗尽有效完成任务所必需的认知资源。①②③ 在这个过程中也会产生情绪,比如,沮丧和焦虑。④⑤ 例如,在经历来自外群体的群体声誉威胁、自身声誉威胁(内群体)和来自内群体的群体声誉威胁时,个体可能会产生预期羞愧和内疚这两种社会情绪,因为在每种情况下,人们都有可能让内群体成员失望。与此相反,预期愤怒可能会在经历自身声誉威胁(外群体)时产生,因为当一个人认为自己受到了不公平的待遇时,就会产生这种情绪。

总之,对于刻板印象威胁效应进行分类可以使我们清晰地认识

① Inzlicht M., McKay L., & Aronson J., "Stigma as ego depletion: How being the target of prejudice affects self-control", *Psychological Science*, Vol. 17, No. 3, May 2006, pp. 262 – 269.

② Johns M., Inzlicht M., & Schmader T., "Stereotype threat and executive resource depletion: Examining the influence of emotion regulation", *Journal of Experimental Psychology-General*, No. 137, December 2008, pp. 691 – 705.

③ McGlone M. S., & Aronson J., "Stereotype threat, identity salience, and spatial reasoning", *Journal of Applied Developmental Psychology*, No. 27, September 2006, pp. 486 – 493.

④ Bosson J. K., Haymovitz E. L., & Pinel E. C., "When saying and doing diverge", *Journal of Experimental Social Psychology*, No. 40, June 2004, pp. 247 – 255.

⑤ Keller J., & Dauenheimer D., "Stereotype threat in the classroom: Dejection mediates the disrupting threat effect on women's math performance", *Personality and Social Psychology Bulletin*, No. 29, March 2003, pp. 371 – 381.

到，这些威胁是由不同的因素引起的。虽然这些刻板印象威胁在消极刻板印象领域的表现影响方面存在相似之处，但区分这些刻板印象威胁具有重要意义，这可以使我们识别刻板印象威胁的风险，从而了解人们如何应对和补偿刻板印象威胁的体验以及在相关领域（如，学校教育）制定相应的干预措施。

四　刻板印象威胁溢出效应

（一）刻板印象威胁溢出效应概述

Steele 和 Aronson 在提出刻板印象威胁效应时指出，当个体认为有关其群体的负面刻板印象将成为评判其行为的一面镜子时会产生相应的消极影响。[①] 当时他们以种族刻板印象为例，指出黑人学生在标准化成绩测试中的表现往往比白人学生差，原因之一就是刻板印象"弥漫在空气中"，引起了他们根深蒂固的恐惧，分散了他们的注意力，使他们无法尽其所能取得好成绩。[②] 另有研究发现，只要暗示一个人的社会身份被贬低和边缘化，当一个人感觉自己是社会身份威胁的受害者时，就会出现这种情况。[③] 但是，当个体脱离和刻板印象有关的环境后，这种消极的影响是否会消失呢？

有研究者提出了刻板印象威胁溢出效应的概念，是指在与威胁来源无关的领域遭受短期和长期影响的社会心理过程。也就是说，

---

① Steele C. M., & Aronson J., "Stereotype threat and the intellectual performance of African Americans", *Journal of Personality and Social Psychology*, No. 69, October 1995, pp. 797 – 811.

② Steele C. M., "A threat in the air: How stereotypes shape intellectual identity and performance", *American Psychologist*, No. 52, May 1997, pp. 613 – 629.

③ Steele C. M., Spencer S. J., & Aronson J., "Contending with group image: The psychology of stereotype and social identity threat", in M. P. Zanna (Ed.), *Advances in Experimental Social Psychology*, San Diego: Academic Press, Vol. 34, 2002, pp. 379 – 440.

刻板印象的中心人物可能会面临各种压力:在短期内,这种压力会促使人们努力应对,消耗做其他事情所需的能量,包括做出正确的决定和调节情绪;从长期来看,这种压力的增加会直接或间接导致身心健康问题,如高血压、肥胖和抑郁。[①]

(二) 刻板印象威胁溢出效应的短期影响

由于应对刻板印象威胁会消耗个体的心理资源,因此有可能影响任何需要自我控制的领域。目前已有多项实验研究证实,在离开威胁环境后,人们会在与最初威胁无关的领域继续表现出不适应行为。比如有研究表明,刻板印象威胁会导致攻击行为、暴饮暴食、冒险决策和注意力不集中。

具体来说,有研究者考察了应对刻板印象威胁是否会导致女性的攻击行为。[②] 在生活中,尽管攻击行为较为常见,但这种行为背后却隐藏着重要的心理机制,其中,自我控制能力显得尤为重要。也就是说,可能是缺乏自我控制能力,导致个体自我耗竭而产生攻击行为,[③][④] 即应对刻板印象威胁会导致自我耗竭,因此也会导致无节制的攻击行为。在这项研究中,女性参与者参加了一次难度很大的数学测试,其中一半人接受了指导,要求她们中立客观地重新

① Pascoe E. A. , & Richman L. , "Perceived discrimination and health: A metaanalytic review", *Psychological Bulletin*, No. 135, March 2009, pp. 531 – 554.

② Inzlicht M. , & Kang S. K. , "Stereotype threat spillover: How coping with threats to social identity affects aggression, eating, decision-making, and attention", *Journal of Personality and Social Psychology*, No. 99, June 2010, pp. 467 – 481.

③ Dewall C. N. , Baumeister R. F. , Stillman T. F. , & Gailliot M. T. , "Violence restrained: Effects of self-regulation and its depletion on aggression", *Journal of Experimental Social Psychology*, No. 43, July 2007, pp. 62 – 76.

④ Stucke T. S. , & Baumeister R. F. , "Ego depletion and aggressive behaviour: Is the inhibition of aggression a limited resource?" *European Journal of Social Psychology*, No. 36, December 2006, pp. 1 – 13.

评估情境和测试。这种重新评估的指导消除了为了应对威胁而压抑思想和情绪的需要，从而节省了参与者的自我控制资源。[①] 另一半被试则没有得到关于如何应对这种情况的进一步指导，她们可能会采取资源消耗的应对策略，这种策略会压抑被试的情绪和认知。[②③]然后，当不再处于威胁情境中时，被试要完成一项与同伴竞争反应时间的任务。在这项任务中，谁对刺激的反应更快，谁就可以向反应较慢的伙伴发出一阵白噪声。实验中评估攻击性的具体表现是向同伴发出白噪声的强度和持续时间。结果发现了刻板印象威胁的溢出效应：应对刻板印象威胁的女性比那些被鼓励重新评估情况的女性做出了更多的攻击行为。也就是说，人们通常会克制自己的攻击冲动，但应对威胁的女性却不会这样。

另一项研究采用同样的实验操作，考察了刻板印象威胁是否会蔓延到饮食行为领域。[④] 与第一项研究一样，女性被试被分为两组，参加一项难度较高的数学测试，一组参与重新评价，另一组不参与重新评价。然后，研究者要求她们参加一个和之前的数学测验无关的"味道测试"，测试三种冰淇淋口味，并允许她们想吃多少就吃多少。实验的逻辑是要克制吃这种令人发胖但诱人的食物的冲动，

---

① Richards J. M., & Gross J. J., "Emotion regulation and memory: The cognitive costs of keeping one's cool", *Journal of Personality and Social Psychology*, No. 79, February 2000, pp. 410 – 424.

② Johns M., Inzlicht M., & Schmader T., "Stereotype threat and executive resource depletion: Examining the influence of emotion regulation", *Journal of Experimental Psychology-General*, No. 137, May 2008, pp. 691 – 705.

③ Logel C., Iserman E. C., Davies P. G., Quinn D. M., & Spencer S. J., "The perils of double consciousness: The role of thought suppression in stereotype threat", *Journal of Experimental Social Psychology*, No. 45, June 2009, pp. 299 – 312.

④ Inzlicht M., & Kang S. K., "Stereotype threat spillover: How coping with threats to social identity affects aggression, eating, decision-making, and attention", *Journal of Personality and Social Psychology*, No. 99, March 2010, pp. 467 – 481.

需要参与者的自我控制资源,因此,自我耗竭的参与者应该不太能够拒绝自己吃冰淇淋。① 结果证实了这一预测:那些可能压抑了自己情绪和想法的被试,也就是威胁组的女性吃冰淇淋的次数明显多于非威胁组的被试。

已有研究表明,自我耗竭会妨碍决策过程中的深思熟虑②,因此,有研究者考察了刻板印象威胁在决策中的效应。在这项研究中,研究者提示被试会在彩票任务中经历一次身份威胁。实验时他们可以在两种彩票中做出选择,一种彩票风险很大,但收益很高;另一种彩票风险小得多,但收益较低。第二种彩票的预期效用更高,因此从理性上讲,第二种低风险彩票是更好的选择。结果再次揭示了一种溢出效应:受到身份威胁的被试受限于自动、直观的决策系统,他们比对照组参与者更常选择风险大的彩票。③

尽管上述研究提供了很好的证据,表明刻板印象威胁会蔓延并影响其他非刻板印象领域的行为,但这些行为实验仅仅能够说明行为的结果,并不能提供产生这些结果的心理机制。为了揭示这些行为背后的神经过程,研究者考察了前扣带回皮层(ACC)的活动④,该大脑区域与大脑边缘和前额叶区域相互联系,对于自我控制至关

---

① Vohs K. D. , & Heatherton T. F. , "Self-regulatory failure: A resource-depletion approach", *Psychological Science*, No. 11, October 2000, pp. 249 – 254.

② Kahneman D. , "A perspective on judgment and choice: Mapping bounded rationality", *American Psychologist*, No. 58, May 2003, pp. 697 – 720.

③ Masicampo E. J. , & Baumeister R. F. , "Toward a physiology of dual-process reasoning and judgment: Lemonade, willpower, and expensive rule-based analysis", *Psychological Science*, No. 19, July 2008, pp. 255 – 260.

④ Inzlicht M. , & Kang S. K. , "Stereotype threat spillover: How coping with threats to social identity affects aggression, eating, decision-making, and attention", *Journal of Personality and Social Psychology*, No. 99, September 2010, pp. 467 – 481.

重要。① 在脑电图研究中，ACC 的激活与内侧额叶负性事件相关电位（ERP）有关，对错误、冲突和不确定性比较敏感。② 这些 ERP是对个人表现的情绪反应的产物③，是由 ACC 发出的神经"求救信号"，它们表明何时需要注意、警惕和控制④。一项研究表明，自我控制耗竭可能是由低效的绩效监控⑤造成的，例如，对错误类型的事件（如，不需要注意或警惕的事件）增加行为监控。如果刻板印象威胁消耗了执行资源，那么它不仅会导致执行控制能力低下，这种影响还应该通过对这种 ACC 绩效监控系统的破坏来调节。

在 Inzlicht 和 Kang 的研究中，男女参与者参加了一项诊断性数学测试，并被要求"自然"应对（控制组）或被鼓励重新评估自己的情绪（实验组）。⑥ 测试结束后，参与者完成了一项旨在挖掘认知抑制过程的 Stroop 颜色命名任务，同时记录他们的 ACC 活动。结果发现，受到威胁的女性参与者在 Stroop 任务中的表现比男性或未受威胁的女性参与者更差。此外，这些对照组参与者的

①　Bush, G. , Luu, P. , & Posner, M. I. , "Cognitive and emotional influences in anterior cingulate cortex", *Trends in Cognitive Sciences*, No. 4, 2000, pp. 215 – 222.

②　Gehring W. J. , Goss B. , Coles M. G. , & Meyer D. E. , "A neural system for error detection and compensation", *Psychological Science*, No. 4, June 1993, pp. 385 – 390.

③　Luu P. , Collins P. , & Tucker D. M. , "Mood, personality and self-monitoring: Negative affect and emotionality in relation to frontal lobe mechanisms of error monitoring", *Journal of Experimental Psychology*: *General*, No. 129, May 2000, pp. 43 – 60.

④　Bartholow B. D. , Pearson M. A. , Dickter C. L. , Fabiani M. , Gratton G. , & Sher K. H. , "Strategic control and medial frontal negativity: Beyond errors and response conflict", *Psychophysiology*, No. 42, November 2005, pp. 33 – 42.

⑤　Inzlicht M. , & Gutsell J. N. , "Running on empty: Neural signals for self-control failure", *Psychological Science*, No. 18, April 2007, pp. 933 – 937.

⑥　Inzlicht M. , & Kang S. K. , "Stereotype threat spillover: How coping with threats to social identity affects aggression, eating, decision-making, and attention", *Journal of Personality and Social Psychology*, No. 99, August 2010, pp. 467 – 481.

大脑活动模式正常，在需要行为抑制的试次后，ERP 的波幅较高；在不需要行为抑制的试次后，ERP 波幅较低。然而，受威胁参与者的 ACC 活动却产生变化：无论是否需要抑制，ERP 波幅都很高；值得注意的是，在不需要抑制的试次之后，ERP 波幅尤其高。因此，受到威胁的参与者似乎在所有试次类型后都会更加警惕和焦虑，并倾向于在不需要警惕的情况下浪费认知资源。由此看来，经历过刻板印象威胁会影响基于 ACC 的行为表现监控系统，使其效率低下，从而影响有效的自我控制。

(三) 刻板印象威胁溢出效应的长期影响

当个体一离开威胁环境，刻板印象威胁的直接后果就会显现出来，除此之外，越来越多的证据表明，这种威胁会对健康产生长期的不利影响。如上所述，感受到对社会身份的明显威胁会使人们感到压力并去努力应对这种压力。这些压力和应对经历会导致生理、心理和行为的变化，从而对健康结果产生显著影响。[1] 许多研究都揭示了感知到的歧视、偏见或不良观念与健康之间的联系，这些研究都发现，心理和生理健康会受到歧视经历的不利影响。[2][3] 因此，刻板印象威胁的溢出效应可以为我们提供一个新的视角来理解这些问题是如何产生的，以及如何避免这些问题。

有大量研究考察了感知到的歧视对心理健康的影响，特别是

---

[1] Pascoe E. A. , & Richman L. , "Perceived discrimination and health: A metaanalytic review", *Psychological Bulletin*, No. 135, May 2009, pp. 531 –554.

[2] Pascoe E. A. , & Richman L. , "Perceived discrimination and health: A metaanalytic review", *Psychological Bulletin*, No. 135, May 2009, pp. 531 –554.

[3] Williams D. R. , & Mohammed S. A. , "Discrimination and racial disparities in health: Evidence and needed research", *Journal of Behavioral Medicine*, No. 32, June 2009, pp. 20 –47.

对抑郁症的影响。比如，对于美国的黑人[1]以及相当多的其他少数群体[2]来说，一个人受到的歧视或偏见越多，他或她就越有可能表现出抑郁症状。此外，一些纵向研究表明，感知到的歧视也会预测抑郁症状的发展。[3]

相关研究表明，焦虑[4][5]、叛逆行为[6]、创伤后应激障碍[7]和幸福感下降[8]等各种心理健康结果都与歧视或偏见有关。一项元分析发现，感知到的歧视与各种心理健康指数之间存在显著的相

① Lincoln K. D., Chatters L. M., Taylor R. J., & Jackson J. S., "Profiles of depressive symptoms among African Americans and Caribbean Blacks", *Social Science & Medicine*, No. 65, July 2007, pp. 200 – 213.

② Williams D. R., & Mohammed S. A., "Discrimination and racial disparities in health: Evidence and needed research", *Journal of Behavioral Medicine*, No. 32, September 2009, pp. 20 – 47.

③ Brody G. H., Chen Y., Murry V. M., Ge X., Simons R. L., Gibbons F. X., et al., "Perceived discrimination and the adjustment of African American youths: A five-year longitudinal analysis with contextual moderation effects", *Child Development*, No. 77, October 2006, pp. 1170 – 1189.

④ Banks K. H., Kohn-Wood L. P., & Spencer M., "An examination of the African American experience of everyday discrimination and symptoms of psychological distress", *Community Mental Health Journal*, No. 42, December 2006, pp. 555 – 570.

⑤ Bhui K., Stansfeld S., McKenzie K., Karlsen S., Nazroo J., & Weich S., "Racial/Ethnic discrimination and common mental disorders among workers: Findings from the EMPIRIC study of ethnic minority groups in the United Kingdom", *American Journal of Public Health*, No. 95, April 2005, pp. 496 – 501.

⑥ Brook J. S., Brook D. W., Balka E. B., & Rosenberg G., "Predictors of rebellious behavior in childhood: Parental drug use, peers, school environment, and child personality", *Journal of Addictive Diseases*, No. 25, March 2006, pp. 77 – 87.

⑦ Khaylis A., Waelde L., & Bruce E., "The role of ethnic identity in the relationship of race-related stress to PTSD symptoms among young adults", *Journal of Trauma & Dissociation*, No. 8, June 2007, pp. 91 – 105.

⑧ Sujoldzic A., Peternel L., Kulenovic T., & Terzic R., "Social determinants of health—a comparative study of Bosnian adolescents in different cultural contexts", *Collegium Antropologicum*, No. 30, May 2006, pp. 703 – 711.

关性。[1] 此外，在对社会经济地位、教育和就业等因素进行控制后，歧视与健康之间的联系依然存在。[2] 虽然该领域的大部分研究都是相关性的，没有涉及因果关系，但纵向研究表明，个体是由于感知到的偏见或歧视影响了心理健康，而不是由于心理不健康导致的偏见或歧视。总之，社会身份威胁会使人们的心理健康和身体健康都受到消极影响。

与心理健康一样，感知到的歧视与身体健康水平下降之间也存在联系。多项研究表明了歧视和偏见导致疾病和风险因素的增加，比如，会导致肥胖[3]、高血压[4][5]，以及自我报告的健康状况不佳[6][7]。一项纵向研究发现，即使在控制乳腺癌风险因素的情况下，

---

① Pascoe E. A. , & Richman L. , "Perceived discrimination and health: A metaanalytic review", *Psychological Bulletin*, No. 135, May 2009, pp. 531 –554.

② Pascoe E. A. , & Richman L. , "Perceived discrimination and health: A metaanalytic review", *Psychological Bulletin*, No. 135, May 2009, pp. 531 –554.

③ Inzlicht, M. , & Kang, S. K. , "Stereotype threat spillover: How coping with threats to social identity affects aggression, eating, decision-making, and attention", *Journal of Personality and Social Psychology*, No. 99, 2010, pp. 467 –481.

④ Davis S. K. , Liu Y. , Quarells R. C. , Din-Dzietharn R. , & M. A. H. D. S. Group, "Stress-related racial discrimination and hypertension likelihood in a population-based sample of African Americans: The Metro Atlanta Heart Disease Study", *Ethnicity and Disease*, Vol. 15, No. 4, October 2005, pp. 585 –593.

⑤ Roberts C. B. , Vines A. I. , Kaufman J. S. , & James S. A. , "Cross-sectional association between perceived discrimination and hypertension in African-American men and women: The Pitt County Study", *American Journal of Epidemiology*, Vol. 167, No. 5, September 2007, pp. 624 –632.

⑥ Harris R. , Tobias M. , Jeffreys M. , Waldegrave K. , Karlsen S. , & Nazroo J. , "Effects of self-reported racial discrimination and deprivation on Māori health and inequalities in New Zealand: Cross-sectional study", *Lancet*, Vol. 367, No. 9527, 2006, pp. 2005 –2009.

⑦ Larson A. , Gillies M. , Howard P. J. , & Coffin J. , "It's enough to make you sick: The impact of racism on the health of Aboriginal Australians", *Australian and New Zealand Journal of Public Health*, Vol. 31, No. 4, April 2007, pp. 322 –329.

那些报告经常受到日常歧视的妇女的乳腺癌发病率也较高。① 那么，这些影响产生的机制是什么？

根据前文所述，人们会对歧视或偏见有两种直接反应：非自愿的压力反应和自愿的应对反应。通过对健康风险因素和健康行为的影响，压力反应和应对反应可以作为感知到的歧视与健康之间关系的中介因素，从而有助于阐明负面健康影响的产生过程。

在考察种族刻板印象的研究中发现，种族主义会增加人的压力，并直接导致生理效应，如血压升高、心率加快和负面情绪反应，而这些都是与压力有关的疾病的指标。② 此外，压力还会对消极情绪等心理因素产生负面影响，从而对心理和身体健康造成不利后果。③

除压力的直接影响之外，个体应对压力的主观努力也会对健康产生间接的负面影响。这些应对策略造成伤害的主要方式之一就是自我耗竭。比如，努力控制思想、情绪和行为会导致个体缺乏完成其他任务所需的资源。④⑤ 由于自我控制能力下降，人们可

① Taylor T. R. , Williams C. D. , Makambi K. H. , Mouton C. , Harrell J. P. , Co-zier Y. , et al. , "Racial discrimination and breast cancer incidence in US black women: The black women's health study", *American Journal of Epidemiology*, No. 166, March 2007, pp. 46 – 54.

② Harrell J. P. , Hall S. , & Taliaferro J. , "Physiological responses to racism and discrimination: An assessment of the evidence", *American Journal of Public Health*, No. 93, November 2003, pp. 243 – 248.

③ Watson D. , & Pennebaker J. W. , "Health complaints, stress, and distress: Exploring the central role of negative affectivity", *Psychological Review*, No. 96, October 1989, pp. 234 – 254.

④ Baumeister R. F. , Faber J. E. , & Wallace H. M. , "Coping and ego-depletion: Recovery after the coping process", in C. R. Snyder (Ed.), *Coping: The Psychology of what Works*, New York: Oxford University Press, 1999, pp. 50 – 69.

⑤ Inzlicht M. , Aronson J. , Good C. , & McKay L. , "A particular resiliency to threatening environments", *Journal of Experimental Social Psychology*, No. 42, May 2006, pp. 323 – 336.

能会表现出较多不健康行为的倾向。[1] 例如，当被试报告感受到较高的歧视或偏见时会产生较多的吸烟[2]、吸毒[3][4][5]和不健康饮食行为[6]，而遵循医疗建议的行为则会减少[7][8]。此外，在实验中引入刻板印象威胁会导致女性吃得更多，这提供了更多证据表明，应对负面刻板印象会导致自我耗竭，从而促使个人做出更不健康的选择。[9]

① Pascoe E. A. , & Richman L. , "Perceived discrimination and health: A metaanalytic review", *Psychological Bulletin*, No. 135, June 2009, pp. 531 –554.

② Landrine H. , & Klonoff E. A. , "The schedule of racist events: A measure of racial discrimination and a study of its negative physical and mental health consequences", *Journal of Black Psychology*, No. 22, March 1996, pp. 144 –168.

③ Martin J. K. , Tuch S. A. , & Roman P. M. , "Problem drinking patterns among African Americans: The impacts of reports of discrimination, perceptions of prejudice, and 'risky' coping strategies", *Journal of Health and Social Behavior. Special Issue: Race, Ethnicity and Mental Health*, No. 44, July 2003, pp. 408 –425.

④ Yen I. H. , Ragland D. R. , Grenier B. A. , & Fisher J. M. , "Workplace discrimination and alcohol consumption: Findings from the San Francisco Muni health and safety study", *Ethnicity & Disease*, Vol. 9, No. 1, October 1999, pp. 70 –80.

⑤ Gibbons F. X. , Yeh H. C. , Gerrard M. , Cleveland M. J. , Cutrona C. , Simons R. L. , & Brody G. H. , "Early experience with racial discrimination and conduct disorder as predictors of subsequent drug use: A critical period analysis", *Drug and Alcohol Dependence*, Vol. 88, No. 1, June 2007, pp. 27 –37.

⑥ Mulia N. , "Social disadvantage, stress, and alcohol use among Black, Hispanic, and White Americans: Findings from the 2005 U. S. national alcohol survey", *Journal of Studies on Alcohol and Drugs*, No. 69, May 2008, p. 824.

⑦ Casagrande S. S. , Gary T. L. , LaVeist T. A. , Gaskin D. J. , & Cooper L. A. , "Perceived discrimination and adherence to medical care in a racially integrated community", *Journal of General Internal Medicine*, Vol. 22, No. 3, July 2007, pp. 389 –395.

⑧ Facione N. C. , & Facione P. A. , "Perceived prejudice in healthcare and women's health protective behavior", *Nursing Research*, No. 56, September 2007, pp. 175 –184.

⑨ Inzlicht M. , & Kang S. K. , "Stereotype threat spillover: How coping with threats to social identity affects aggression, eating, decision-making, and attention", *Journal of Personality and Social Psychology*, No. 99, February 2010, pp. 467 –481.

五　性别刻板印象威胁效应相关研究

种族刻板印象研究领域的研究者最早发现了刻板印象威胁效应，他们的研究考察了种族和智力刻板印象背景下的刻板印象威胁，因为在文化上存在着黑人不如白人聪明的刻板印象。[①] 在这项研究中，一所名牌大学的黑人和白人大学生参加了一项难度很大的口语考试，考试项目取自研究生入学考试（GRE）。有一半的考生被告知这次考试与智商测验一样，是用来衡量他们的智力能力的。这种"诊断性"条件的目的是制造出人们在智力受到审查时所面临的心理状况。将参与者的表现与"非诊断性"条件下的测试者的表现进行比较，在"非诊断性"条件下，研究者告诉测试者这项研究与智力无关，他们的能力不会受到评估。设计"非诊断性"条件的目的是尽量减少非裔美国人智力较低这一刻板印象的相关影响。在其他各方面，条件都是相同的。这项研究结果表明，虽然白人学生不受诊断性操纵的影响，但与非诊断性测验相比，当测验被定为智力诊断性测验时，黑人学生的成绩要差得多。在一项后续研究中，研究者发现，仅仅要求黑人受试者参加测试前在问卷上标明自己的种族，就足以诱发刻板印象威胁，并影响他们在其他非诊断性（也可能是非威胁性）情境中的表现；被要求标明种族的黑人学生解决的题目数量大约是未被要求标明种族的黑人学生的一半。然而，白人学生在参加测试前并没有受到其种族识别的影响。

---

① Steele C. M. , & Aronson J. , "Stereotype threat and the intellectual performance of African Americans", *Journal of Personality and Social Psychology*, No. 69, April 1995, pp. 797 – 811.

  Spencer 及其同事在数学领域进行了类似的研究。[1] 在这些研究中,研究人员关注的重点是参与者的性别,因为在相关的刻板印象中,女性在数学和科学方面往往不如男性。在刻板印象威胁条件下,实验者在进行测试前做了一个简单的陈述,告诉参与者过去男性和女性在测试中的表现是不同的。在控制条件下,实验者的陈述是,过去的测试并没有显示出性别差异。在刻板印象威胁条件下,男性的表现明显优于女性。然而,当在控制条件下消除刻板印象威胁,即向女性保证测试是性别公平时,女性的表现与男性相似。

  有关刻板印象威胁的早期研究表明,刻板印象威胁的体验可能有所不同,这取决于群体成员避免刻板印象的动机如何,以及采用何种形式避免刻板印象。比如,与人文学科相比,接触过刻板印象广告的女性对数学和科学的职业兴趣较低。[2] 同样,观看了有关性别不平衡的会议视频的女性比男性或观看了性别比例平衡视频的女性更不愿意参加会议。[3] 目标也可以通过尽量不去想刻板印象来避免刻板印象,也就是说,开始进行高威胁数学测试的女性会抑制对刻板印象的思考。[4]

---

[1] Spencer S. J. , Steele C. M. , & Quinn D. M. , "Stereotype threat and women's math performance", *Journal of Experimental Social Psychology*, No. 35, August 1999, pp. 4 – 28.

[2] Davies P. G. , Spencer S. J. , Quinn D. M. , & Gerhardstein R. , "Consuming images: How television commercials that elicit stereotype threat can restrain women academically and professionally", *Personality and Social Psychology Bulletin*, No. 28, December 2002, pp. 1615 – 1628.

[3] Murphy M. C. , Steele C. M. , & Gross J. J. , "Signaling threat: How situational cues affect women in math, science, and engineering settings", *Psychological Science*, No. 18, May 2007, pp. 879 – 885.

[4] Logel C. , Iserman E. C. , Davies P. G. , Quinn D. M. , & Spencer S. J. , "The perils of double consciousness: The role of thought suppression in stereotype threat", *Journal of Experimental Social Psychology*, No. 45, March 2009, pp. 299 – 312.

　　上述研究以及随后的大量研究一致表明，一些简单的情境因素，如得知某项测试是对某项刻板能力的诊断、在考试前对人口统计部分进行标记、在特定情境中知道自己是少数群体成员之一、看到对某群体成员的刻板描述、认为评价者有偏见（如，性别歧视或种族主义）等，都会出现刻板印象威胁导致的成绩下降。[1][2][3][4]

六　刻板印象威胁效应的应对与干预

　　尽管当人们处于消极刻板印象的中心，并表现出行为或认知方面的损伤，但人们可能会以不同的方式应对每一种刻板印象威胁。[5] 相关研究表明，刻板印象威胁会导致行为表现下降，或为自己的成功设置破坏性障碍[6][7]，脱离或贬低刻板印

①　Davie P. G. , Spencer S. J. , & Steele C. M. , "Clearing the air: Identity safety moderates the effects of stereotype threat on women's leadership aspirations", *Journal of Personality and Social Psychology*, No. 88, June 2005, pp. 276 – 287.

②　Nguyen H. D. , & Ryan A. M. , "Does stereotype threat affect test performance of minorities and women? A meta-analysis of experimental evidence", *Journal of Applied Psychology*, No. 93, October 2008, pp. 1314 – 1334.

③　Walton G. M. , & Cohen G. L. , "A question of belonging: Race, social fit, and achievement", *Journal of Personality and Social Psychology*, No. 92, November 2007, pp. 82 – 96.

④　Walton G. , Spencer S. J. , & Erman S. , Affirmative meritocracy, Manuscript under review, 2011.

⑤　Shapiro J. R. , & Neuberg S. L. , "From stereotype threat to stereotype threats: Implications of a multi-threat framework for causes, moderators, mediators, consequences, and interventions", *Personality and Social Psychology Review*, No. 11, December 2007, pp. 107 – 130.

⑥　Keller J. , "Blatant stereotype threat and women's math performance: Self-handicapping as a strategic means to cope with obtrusive negative performance expectations", *Sex Roles*, No. 47, June 2022, pp. 193 – 198.

⑦　Stone J. , "Battling doubt by avoiding practice: The effects of stereotype threat on self-handicapping in White athletes", *Personality and Social Psychology Bulletin*, No. 28, September 2002, pp. 1667 – 1678.

象领域[1][2]，以及降低追求与刻板印象相关职业的愿望[3]。尽管在面对刻板印象威胁时，诸如反对刻板印象等的策略可能会比较常用（例如，反对女性不擅长数学的观点），但还是要考虑认知和行为因素，这样才能在面对刻板印象威胁时减少其负面影响。

举例来说，来源于自我的刻板印象威胁侧重于一个人对自我的看法（自我概念威胁）和对群体的看法（群体概念威胁），因此很可能采取旨在维护积极的自我概念或群体观点的应对策略（例如，自我打击、忽视反馈、制造借口、贬低该领域等）。与此相反，来源于他人的刻板印象威胁则侧重于群体内和群体外他人对自我或群体的看法，因此往往需要采取公开的补救策略（例如，公开自我打击、公开制造借口）。需要注意的是，来自他人的刻板印象威胁可能会引起愤怒，当一个人认为自己受到不公平对待时，往往会出现这种情绪反应，而来自自身的刻板印象威胁则可能会引起羞愧。因此，这些情绪反应会激发不同类型的策略。例如，在个人认为自己拥有足够资源的情况下（如，非常有才华的女性工程师），愤怒可能更容易引发挑战反应，从而促进

① Major B. , & Schmader T. , "Coping with stigma through psychological disengagement", in J. Swim, & C. Stangor (Eds.), *Prejudice: The Target's Perspective*, San Diego: Academic Press, 1998.

② Nussbaum D. A. , & Steele C. M. , "Situational disengagement and persistence in the face of adversity", *Journal of Experimental Social Psychology*, No. 43, April 2007, pp. 127 – 134.

③ Davies P. G. , Spencer S. J. , Quinn D. M. , & Gerhardstein R. , "Consuming images: How television commercials that elicit stereotype threat can restrain women academically and professionally", *Personality and Social Psychology Bulletin*, No. 28, May 2002, pp. 1615 – 1628.

更成功的表现。①②

在应对来自群体目标的刻板印象威胁时，人们可能会采取一些策略来减少对群体的依恋：比如，通过降低群体本身与个人自我价值的相关程度。脱离群体可以减少个人在与刻板印象相关的表现之前可能会产生的不适感。一个人还可以通过直接疏远作为群体代表的自己来应对（如，指出其他群体成员更适合作为代表的角色，传达有关自己或其他群体成员成功的补偿信息）。对于以自我为目标的刻板印象威胁，一个人也可能会拒绝或以其他方式与被刻板印象的群体保持距离，试图摆脱以群体的刻板印象能力来评价自己的可能性。然而，鉴于对群体的认同并不会促成自我即目标刻板印象威胁的体验，体验到自我即目标刻板印象威胁的个体也可能会参与群体，或拉近自己与群体的距离，以此作为支持的来源。③

由于刻板印象威胁效应可能对个体造成认知、情绪和行为表现的下降，因此，研究者开始关注对这种威胁效应的干预。由于刻板印象威胁效应的类型不同，因此，有效缓解一种刻板印象威胁的干预措施不太可能有效缓解另一种刻板印象威胁。这表明，干预措施要想有效，就必须针对破坏认知或行为表现，或引发有害应对策略的特定刻板印象威胁来进行。

---

① Crisp R. J., Bache L. M., & Maitner A. T., "Dynamics of social comparison in counterstereotypic domains: Stereotype boost, not stereotype threat, for women engineering majors", *Social Influence*, No. 4, June 2009, pp. 171 – 184.

② Kray L. J., Reb J., Galinsky A. D., & Thomson L., "Stereotype reactance at the bargaining table: The effect of stereotype activation and power on claiming and creating value", *Personality and Social Psychology Bulletin*, No. 30, October 2004, pp. 399 – 411.

③ Crocker J., Major B., & Steele C. M., "Social stigma", in D. Gilbert, S. T. Fiske, & G. Lindzey (Eds.), *The Handbook of Social Psychology* (4th ed.), Boston: McGraw Hill, Vol. 2, 1998, pp. 504 – 553.

　　首先，只要我们能确定个人或群体所经历的刻板印象威胁的具体形式，我们就能针对最相关的诱发情况制定干预措施。其次，如果希望通过一次干预来消除几种形式的刻板印象威胁，那么干预就需要多管齐下：需要改变几组特定的条件。例如，如果我们能让一个人相信，评估她数学成绩的人永远不会知道她的身份，那么她就不应该经历刻板印象威胁。也就是说，即使存在引发这种刻板印象威胁的其他条件，即她知道别人知道她的性别；她知道别人认为女性缺乏数学天赋；她非常希望别人不要看到她数学成绩差等，如果她确信她的身份不会与她的成绩联系在一起，她也不会体验到这种威胁。这并不是说她不会遭遇其他形式的刻板印象威胁；她很可能还会遭受其他威胁效应，因为产生威胁效应的因素多种多样。因此，研究表明，并不存在单一、简单的干预措施：不同的干预措施需要针对引发每种刻板印象威胁的不同因素或每种刻板印象威胁的不同机制来进行。

第二章

性别刻板印象与儿童发展的
理论与实证研究

第一节　儿童刻板印象发展的相关理论

基于人类性别对事物进行分类是一种基本现象，几乎影响到人们日常生活的方方面面。以性别角色发展和功能为基础的相关社会认知理论阐明了性别观念是如何从复杂的经验组合中建构起来的，以及它们是如何与动机和自我调节机制协同作用，在整个生命过程中指导与性别相关的行为的。这些理论在一个统一的概念结构中整合了心理和社会结构决定因素。这些理论告诉我们，性别观念和性别角色是在各种社会子系统中相互依存的广泛社会影响网络的产物。人类的进化提供了身体结构和生物潜能，产生出一系列的可能性，而不是规定一种固定的性别分化类型。人们通过在相互关联的影响系统中采取积极行动，促进自我发展，并带来界定和构建性别关系的社会变革。

性别发展是一个在生活中无法避免的问题，因为人们生活中最重要的一些方面，如他们培养的才能、他们对自己和他人的观念、他们所遇到的机会和限制，以及他们所追求的社会生活和职业道路，在很大程度上都是由社会性别类型所规定的。这是人们被区分的主要依据，对人们的日常生活产生了普遍影响。性别差异之所以更加重要，是因为男性和女性的许多特质和角色往往受到不同的重视。以往一些文献指出，男性的特质和角色通常被认为更受欢迎、地位更高。① 虽然有些性别差异是由生物因素造成的，但大多数与性别相关的刻板化特征和角色更多的是来自文化，而非生物本身的生理属性。②③④

多年来，人们提出了几种解释性别发展的主要理论。这些理论在几个重要方面存在区别：第一个方面涉及对心理、生物和社会结构决定因素的相对重视。以心理学为导向的理论倾向于强调性别发展的内在心理过程。⑤⑥ 相比之下，社会学理论侧重于性别角色发展和功能的社会结构决定因素。⑦⑧⑨ 根据以生物学为导向的理论，性别角色发展和分化的基础是男性和女性在生殖过程中

① Berscheid E., "Forward", in A. E. Beall & R. J. Sternberg (Eds.), *The Psychology of Gender*, New York: Guilford Press, 1993.

② Bandura A., *Social Foundations of Thought and Action: A Social Cogntive Theory*, Englewood Cliffs, NJ: Prentice-Hall, 1986.

③ Beall A. E., & Sternberg R. J. (Eds.)., *The Psychology of Gender*, New York: Guilford Press, 1993.

④ Epstein C. F., "The multiple realities of sameness and difference: Ideology and practice", *Journal of Social Issues*, No. 53, May 1997, pp. 259-278.

⑤ Freud S., *Three Contributions to the Theory of Sex*, New York: Nervous and Mental Disease Publishing Co. (original work published 1905), pp. 1905 – 1930.

⑥ Kohlberg L., "A cognitive-developmental analysis of children's sex-role concepts and attitudes", In E. E. Maccoby (Ed.), *The Development of Sex Differences*, Stanford, CA: Stanford University Press, 1966, pp. 82 – 173.

⑦ Eagly A. H., *Sex Differences in Social Behavior: A Social Role Interpretation*, Hillsdale, NJ: Erlbaum, 1987.

⑧ Berger J., Rosenholtz S. J., & Zelditch, M., "Status organizing processes", *Annual Review of Sociology*, No. 6, August 1980, pp. 479 – 508.

⑨ Epstein C. F., *Deceptive Distinctions: Sex, Gender, and the Social Order*, New Haven, CT: Yale University Press, 1988.

所扮演的不同生物角色所产生的性别差异。[1][2]

　　第二个方面涉及传播模式的性质。心理学理论通常强调在家庭传播模式中对性别观念和行为方式的认知建构。弗洛伊德强调通过认同过程在家庭中采纳性别角色；行为主义理论也重视父母在塑造和规范与性别有关的行为方面的作用。在倾向于生物决定因素的理论中，家庭基因被认为是性别分化的跨代传播媒介。[3]以社会学为导向的理论则强调，性别角色主要是在制度层面上的社会建构。[4]性别角色发展和功能的社会认知理论在一个统一的概念框架内整合了心理和社会结构的决定因素。[5]从这一角度看，性别观念和角色行为是广泛的社会影响网络的产物，这些影响既来自家庭，也来自日常生活中遇到的许多社会系统。[6]因此，它倾向于多方面的社会传播模式，而不是主要的家庭传播模式。

　　第三个方面涉及理论分析的时间范围。大多数心理学理论将性别发展主要视为幼儿期的现象，而不是贯穿整个生命过程的现象。然而，在不同的社会环境和人生的不同时期，性别角色的行为规则在某种程度上是不同的。此外，由于社会文化和技术的变化，有必要对以前存在的关于什么是适当的性别行为的观念进行修正。性别

---

[1]　Buss D. M. , "Psychological sex differences: Origins through sexual selection", *American Psychologist*, No. 50, March 1985, pp. 164 – 168.

[2]　Trivers R. L. , "Parental investment and sexual selection", In B. Campbell (Ed.), *Sexual Selection and the Descent of Man* 1871 – 1971, Chicago: Aldine, 1972, pp. 136 – 172.

[3]　Rowe D. C. , *The Limits of Family Influence: Genes, Experience, and Behavior*, New York: The Guilford Press, 1994.

[4]　Lorber J. , *Paradoxes of Gender*, New Haven, CT: Yale University Press, 1994.

[5]　Bandura A. , *Social Foundations of Thought and Action: A Social Cogntive Theory*, Englewood Cliffs, NJ: Prentice-Hall: 1986.

[6]　Bandura A. , *Self-efficacy: The Exercise of Control*, New York: W. H. Freeman, 1997.

角色的发展和功能的发挥并不局限于童年时期,而是在整个生命过程中都要经过协商。大多数性别发展理论都关注早期发展①②或关注成年人③,而社会认知理论则从生命过程的角度出发。

一　精神分析理论

精神分析理论提出了男孩和女孩性别发展的不同过程。最初,人们认为,男孩和女孩都认同母亲。然而,在3—5岁期间,这种情况会发生变化,儿童会认同同性父母。

在认同的过程中,孩子们会全盘接受同性父母的特征和品质。通过这一认同过程,儿童成为性别类型的人。由于男孩对同性父母的认同比女孩更强烈,因此男孩的性别类型也会更强烈。尽管精神分析理论在早期对发展心理学产生了广泛的影响,但几乎没有经验证据支持这一理论。与同性父母的认同和性别角色采纳之间的明确关系从未得到经验验证。④⑤⑥ 儿童更有可能以养育型或社会强势型

---

① Freud S. , "Introductory lectures on psychoanalysis", in J. Strachey (Ed.), *The Standard Edition of the Complete Psychological Works of Sigmund Frued*, London: Hogarth (original work published 1916): 1916/1963.

② Kohlberg L. , "A cognitive-developmental analysis of children's sex-role concepts and attitudes", in E. E. Maccoby (Ed.), *The Development of Sex Differences*, Stanford, CA: Stanford University Press, 1966, pp. 82 – 173.

③ Deaux K. , & Major B. , "Putting gender into context: An interactive model of gender related behavior", *Psychological Review*, No. 94, May 1987, pp. 369 – 389.

④ Hetherington E. M. , "The effects of familial variables on sex typing, on parent-child similarity, and on imitation in children", in J. P. Hill (Ed.), *Minnesota Symposia on Child Psychology*, Minneapolis: University of Minnesota Press, Vol. 1, 1967, pp. 82 – 107.

⑤ Kagan J. , "The acquisition and significance of sex-typing and sex-role identity", in M. Hoffman & L. Hoffman (Eds.), *Review of Child Development Research*, New York: Russell Sage, Vol. 1, 1964, pp. 137 – 167.

⑥ Payne D. E. , & Mussen, P. H. , "Parent-child relations and father identification among adolescent boys", *Journal of Abnormal and Social Psychology*, No. 52, April 1956, pp. 358 – 362.

为行为榜样，而不是以与他们有竞争关系的威胁型为行为榜样。①

二　认知发展理论

根据认知发展理论，性别认同被认为是儿童性别学习的基本组织者和调节者。② 儿童从周围的所见所闻中形成对性别的刻板化观念。一旦他们实现了性别恒定，也即相信自己的性别是固定的、不可逆转的，他们就会积极地珍视自己的性别身份，并只寻求与这一概念相一致的行为方式。由于发展出认知一致性，个体会试图以与自我认知一致的方式行事。Kohlberg 提出了以下认知过程来创造和维持这种一致性，比如，男孩可能会认为，"我是男孩，所以我想做男孩的事，因此做男孩的事（并因做男孩的事而获得认可）是有回报的"③。根据这种观点，儿童的许多行为都是为了确认自己的性别身份。一旦儿童对自己的性别有了认识，他们的行为（表现得像女孩）和想法（我是女孩）之间的相互影响就会导致稳定的性别认同，或者用认知发展理论的术语来说，儿童实现了性别恒定性。

Kohlberg 将性别恒定性定义为：一个人认识到自己的性别是与基本生物属性相联系的永久属性，而不取决于头发长度、服装款式或游戏活动选择等表面特征。④ 性别恒定性的发展并不是一种全或

① Bandura A. , Ross D. , & Ross S. A. , "A comparative test of the status envy, social power, and secondary reinforcement theories of identificatory learning", *Journal of Abnormal and Social Psychology*, No. 67, August 1963, pp. 527 – 534.

② Kohlberg L. , "A cognitive—developmental analysis of children's sex-role concepts and attitudes", In E. E. Maccoby (Ed.), *The Development of Sex Differences*, Stanford, CA: Stanford University Press, 1966, pp. 82 – 173.

③ Kohlberg L. , "A cognitive—developmental analysis of children's sex-role concepts and attitudes", In E. E. Maccoby (Ed.), *The Development of Sex Differences*, Stanford, CA: Stanford University Press, 1966, pp. 82 – 173.

④ Kohlberg L. , "A cognitive—developmental analysis of children's sex-role concepts and attitudes", In E. E. Maccoby (Ed.), *The Development of Sex Differences*, Stanford, CA: Stanford University Press, 1966, pp. 82 – 173.

无的现象，它包括三个不同层次的性别理解。① 从最不成熟到最成熟的性别理解形式，分别称为性别恒常性的性别认同、性别稳定性和性别一致性。性别认同要求能够简单地将自己标记为男孩或女孩，将他人标记为男孩、女孩、男人或女人。性别稳定性，是指认识到性别会随着时间的推移而保持不变，也就是说，一个人现在的性别和他还是婴儿时的性别是一样的，成年后也会保持不变。性别恒定性的最后一个组成部分是性别一致性，大约在六七岁时掌握，此时孩子已经掌握了更多的知识，即尽管外貌、衣着或活动发生了变化，但性别是不变的。儿童通常要到六岁左右才会把自己不可改变地视为男孩或女孩，在此之前，他们不会始终如一地采取性别类型的行为。

三　性别图式理论

为了解释性别发展和性别分化，人们提出了几种性别模式理论。Bem 和 Markus 及其同事提出的社会心理学方法主要集中在性别图式处理信息的个体差异上。②③ Martin 和 Halverson 的方法强调图式发展和功能发展方面。④ 这一理论与认知发展理论有许多相似之处，但在几个方面与认知发展理论有所不同。它不要求性别取向的发展必须达到性别恒定，而是认为只有掌握了性别认同，即儿童

---

① Slaby R. G. , & Frey K. S. , "Development of gender constancy and selective attention to same-sex models", *Child Development*, No. 46, June 1975, pp. 849 – 856.

② Bem S. L. , "Gender schema theory: A cognitive account of sex typing", *Psychological Review*, No. 88, June 1981, pp. 354 – 364.

③ Markus H. , Crane M. , Bernstein S. , & Siladi M. , "Self-schemas and gender", *Journal of Personality and Social Psychology*, No. 42, October 1982, pp. 38 – 50.

④ Martin C. L. , & Halverson C. F. , "A schematic processing model of sex typing and stereotyping in children", *Child Development*, No. 52, March 1981, pp. 1119 – 1134.

能够将自己和他人标记为男性或女性，才能开始性别模式的发展。①
这种模式一旦形成，就会扩展到包括活动和兴趣的知识、个性和社
会属性，以及与性别有关的活动的脚本。②③④ 图式可能是在与环境
的互动中形成的，但构成图式知识结构的性别特征的抽象过程仍未
确定。

一旦形成这种模式，儿童的行为就会与传统的性别角色相一
致。在认知发展理论中，指导儿童与性别相关行为的动力依赖于性
别标签匹配，儿童希望与其他同性一样。例如，洋娃娃被贴上"女
孩用"和"我是女孩"的标签，这意味着"洋娃娃是我用的"。⑤

经验测试的结果使人们对性别模式的决定性作用产生怀疑。
将性别标签与活动和同伴偏好联系起来的相关研究结果也较为混
杂。少数研究发现了两者之间的联系，其他一些研究则报告了在
不同的性别相关行为测量中相互矛盾的结果⑥，还有一些研究根
本没有发现任何联系⑦。即使在那些报告了这种关系的研究中，
性别标签和与性别有关的偏好是因果关系还是仅仅是社会影响和

①　Martin C. L. , & Halverson C. F. , "A schematic processing model of sex typing and stereotyping in children", *Child Development*, No. 52, June 1981, pp. 1119 – 1134.

②　Levy G. D. , & Fivush R. , "Scripts and gender: A new approach for examining gender-role development", *Developmental Review*, No. 13, July 1993, pp. 126 – 146.

③　Martin C. L. , "Stereotypes about children with traditional and nontraditional gender roles", *Sex Roles*, No. 33, August 1995, pp. 727 – 751.

④　Martin C. L. , & Halverson C. F. , "A schematic processing model of sex typing and stereotyping in children", *Child Development*, No. 52, April 1981, pp. 1119 – 1134.

⑤　Martin C. L. , & Halverson C. F. , "A schematic processing model of sex typing and stereotyping in children", *Child Development*, No. 52, April 1981, pp. 1119 – 1134.

⑥　Martin C. L. , & Little J. K. , "The relation of gender understanding to children's sex-typed preferences and gender stereotypes", *Child Development*, No. 61, February 1990, pp. 1427 – 1439.

⑦　Fagot B. I. , "Changes in thinking about early sex role development", *Developmental Review*, No. 5, October 1985, p. 8398.

认知能力的共同作用，仍有待确定。[①] 对与性别有关的行为做出评价性反应的父母，其子女很早就会被贴上性别标签。[②] 因此，性别标签和性别偏好可能都是父母影响的产物。

对性别刻板印象的了解，也就是对男性和女性属性的一般先入之见，同样与性别相关行为无关。[③④⑤] 儿童对性别活动的偏好在他们知道这些活动的性别联系之前就已经出现了。[⑥⑦⑧⑨] 性别模式代表了一种关于男性和女性的更普遍的知识结构。性别图式理论预测，儿童拥有的性别知识越丰富，他们表现出的与性别相关的偏好就越强烈。然而，这种假设的关系并没有得到经验上的

---

① Fagot B. I. , & Leinbach M. D. , "The young child's gender schema: Environmental input, internal organization", *Child Development*, No. 60, May 1989, pp. 663 - 672.

② Fagot B. I. , & Leinbach M. D. , "The young child's gender schema: Environmental input, internal organization", *Child Development*, No. 60, May 1989, pp. 663 - 672.

③ Huston A. C. , "Sex typing", in P. H. Mussen (Series Ed.) & E. M. Hetherington (Vol. Ed.), *Handbook of Child Psychology: Vol. 4. Socialization, Personality, and Social Development* (4th ed.), New York: Wiley, 1983, pp. 387 - 467.

④ Martin C. L. , "New directions for investigating children's gender knowledge", *Developmental Review*, No. 13, March 1993, pp. 184 - 204.

⑤ Signorella M. L. , "Gender schemata: Individual differences and context effects", in L. S. Liben & M. L. Signorella (Eds.), *Children's Gender Schemata: New Directions for Child Development*, San Francisco, CA: Jossey-Bass, Vol. 38, 1987, pp. 23 - 37.

⑥ Blakemore J. E. O. , & Larue A. A. , & Olejnik A. B. , "Sex-appropriate toy preferences and the ability to conceptualize toys as sex-role related", *Developmental Psychology*, No. 15, August 1979, pp. 339 - 340.

⑦ Martin C. L. , "New directions for investigating children's gender knowledge", *Developmental Review*, No. 13, March 1993, pp. 184 - 204.

⑧ Perry D. G. , White A. J. , & Perry L. C. , "Does early sex typing result from children's attempts to match their behavior to sex role stereotypes?" *Child Development*, No. 55, September 1984, pp. 2114 - 2121.

⑨ Weinraub M. , Clemens L. P. , Sockloff A. , Ethridge T. , Gracely E. , & Myers, B. , "The development of sex role stereotypes in the third year: Relationships to gender labeling, gender identity, sex-typed toy preference, and family characteristics", *Child Development*, No. 55, June 1984, pp. 1493 - 1503.

支持。[1] 例如，成年人可能完全了解性别刻板印象，但这并不会随着性别知识的增加而产生与性别相关的行为预测。这些不同的结果未能证实，性别知识是性别相关行为的决定因素。

性别图式理论为研究性别模式形成后对性别信息的认知处理提供了一个有用的框架，它揭示了性别图式处理如何影响与性别相关信息的注意、组织和记忆。[2][3] 其他以成人为研究对象的性别模式也同样证明了信息加工中的性别偏差。[4][5] 性别图式越突出或可用性越强，个体就越会关注、编码、表征和检索与性别相关的信息。然而，性别图式处理与儿童或成人的性别行为无关，或者说，不同的性别图式化测量结果不一致。

四 生物学理论

也有人提出以生物学为导向的理论来解释性别发展和分化。进化心理学就是这样一种理论，它认为，性别分化是祖先设定的。[6][7] 该理论从配偶偏好、生殖策略、父母对后代的投资以及男性的攻击

---

① Martin C. L. , "Children's use of gender-related information in making social judgments", *Developmental Psychology*, No. 25, July 1991, pp. 80 – 88.

② Carter D. B. , & Levy G. D. , "Cognitive aspects of children's early sex-role development: The influence of gender schemas on preschoolers memories and preferences for sex-typed toys and activities", *Child Development*, No. 59, May 1988, pp. 782 – 793.

③ Ruble D. , & Martin C. , "Gender development", in W. Damon (Ed.), *Handbook of Child Psychology* (5th ed.), New York: Wiley, 1998, pp. 933 – 1016.

④ Bem S. L. , "Gender schema theory: A cognitive account of sex typing", *Psychological Review*, No. 88, July 1981, pp. 354 – 364.

⑤ Markus H. , Crane M. , Bernstein S. , & Siladi M. , "Self-schemas and gender", *Journal of Personality and Social Psychology*, No. 42, March 1982, pp. 38 – 50.

⑥ Archer J. , "Sex differences in social behavior: Are the social role and evolutionary explanations compatible?" *American Psychologist*, No. 51, December 1996, pp. 909 – 917.

⑦ Simpson J. A. , & Kenrick D. T. (Eds.). , *Evolutionary Social Psychology*, Mahwah, NJ: Lawrence Erlbaum Associates, 1997.

性等方面分析了性别角色差异的起源。从这个角度看,当代的性别差异起源于祖先对男女面临的不同生殖需求的成功适应。男性对后代生存机会的贡献较小,因此他们寻求多个伴侣,对交配对象的选择也不那么挑剔。此外,父子关系的不确定性增加了在非亲生子女身上投入资源的风险。与此相反,妇女必须怀胎十月,并在孩子出生多年后照顾他们的后代。为了适应她们在生育和养育子女方面所扮演的更重要的角色,女性倾向于选择较少的性伴侣,并倾向于选择那些能够长期为自己和后代提供基本生活必需品的人。与此相反,男性则试图通过与众多年轻且身体迷人的女性繁衍后代来最大限度地提高父子关系的可能性,这表明他们的生育能力很高。由于体型和力量上的优势,雄性通过对雌性行使攻击性统治来解决因生殖利益冲突而产生的问题。强制性力量使雄性能够控制雌性的性行为,并与许多雌性交配。[1][2] 作为这种进化史的遗产,女性比男性更多地将精力投入养育子女的角色中。反过来,男性也进化成了攻击者、社会统治者,因为这种行为增加了他们传播基因的成功率。根据进化心理学,目前许多性别差异,如偏好的性伴侣数量、选择性伴侣的标准、攻击性、嫉妒以及他们所扮演的角色,都源于祖先性别差异的生殖策略。[3] 例如,研究发现,男性更喜欢年轻、身体有吸引力的女性,而女性更喜欢经济条件好的男性作为配偶,这被认为是生物选择的支持。

---

① Smuts B., "Male aggression against women: An evolutionary perspective", *Human Nature*, No. 3, May 1992, pp. 1 – 44.

② Smuts B., "The evolutionary origins of patriarchy", *Human Nature*, No. 6, April 1995, pp. 1 – 32.

③ Buss D. M., & Schmitt D. P., "Sexual strategies theory: An evolutionary perspective on human mating", *Psychological Review*, No. 100, October 1993, pp. 204 – 232.

五　社会学理论

在社会学理论中，性别是一种社会建构，而非生物学上的既定事实。性别差异的根源更多在于社会和制度实践，而非个人的固定属性。Geis 利用各种研究成果，巧妙地记录了性别差异的社会建构和陈规定型观念的长期存在。[①] 性别刻板印象以选择性的性别方式影响着人们对男性和女性的看法、评价和待遇，从而产生了证实最初刻板印象的行为模式。社会行为中的许多性别差异被视为两性分工的产物，这些分工通过受不同性别地位和权力支配的社会结构实践得以复制。[②]

许多社会学家反对将性别一分为二的观点，认为男女在思维和行为方式上的相似性远远超过了他们之间的差异。[③] 随着机会结构和限制性制度安排方面的社会变革，性别差异随着时间的推移逐渐减小。[④] 性别并不是一个单一的整体。同质化的性别类型忽略了女性之间的巨大差异，以及男性因社会经济阶层、教育、种族和职业不同而产生的类似的巨大差异。将所有男性和女性归入二分法性别类别的做法，即男性承担代理职能，女性承担表达和交流职能的做法，同样受到严厉批评。关于情感刻板印象，Epstein 提出，尽管女性理应比男性更易动情，但在伊朗等中东地区的文化

①　Geis F. L. ，"Self-fulfilling prophecies: A social psychological view of gender", in A. E. Beall & R. J. Sternberg (Eds.), *The Psychology of Gender*, New York: Guilford Press, 1993, pp. 9 – 54.

②　Eagly A. H. , *Sex Differences in Social Behavior: A Social Role Interpretation*, Hillsdale, NJ: Erlbaum, 1987.

③　Gerson K. , "Continuing controversies in the sociology of gender", *Sociological Forum*, No. 5, April 1990, pp. 301 – 310.

④　Eagly A. H. , "Reporting sex differences", *American Psychologist*, No. 42, June 1987, pp. 755 – 756.

中，表达情感最强烈的却是男性。① 她认为，那些认为男性和女性在心理构成上存在本质区别的性别理论家正在助长性别刻板印象和两极分化。②

第二节 儿童关于特质的性别差异信念的发展 ——实证研究

一 前言

从出生开始，个体就被纳入两性范畴。除了生理性别的不同，人们的社会性别也被赋予不同的描述。比如，人们通常认为，男性强壮、外向，充满智慧并具有社交能力，而女性则被描述为敏感、内向，爱猜疑，温柔并充满爱心。并且，早期研究表明，男性的特质被简要概括为强调工具性活动，而女性的特质则被概括成强调情感的表达。③④

然而，以上研究均以成人为被试，其研究结果反映的是成人眼中男女性特质的差异，那么在儿童眼中，特质是否存在性别差异呢？或者说，儿童是否持有关于特质的性别差异信念呢？一项以美国、爱尔兰、英格兰儿童为被试的研究发现，关于特质的性别差异

---

① Epstein C. F. , *Deceptive Distinctions*: *Sex*, *Gender*, *and the Social Order*, New Haven, CT: Yale University Press, 1988.

② Gilligan C. , *In a Different Voice*, Cambridge: Harvard University Press, 1982.

③ Rosenkrantz P. S. , Vogel S. R. , Bee H. L. , Broverman I. K. , & Broverman D. M. , "Sex-role stereotypes and self-concepts in college students", *Journal of Consulting and Clinical Psychology*, Vol. 32, No. 3, November 1968, pp. 287 – 295.

④ Williams J. E. , & Bennett S. , "The definition of sex stereotypes via the adjective check list", *Sex Roles*, Vol. 1, No. 4, May 1975, pp. 327 – 337.

信念在 5—11 岁才开始发展；① Ruble 等人认为，5 岁之前的儿童基本没有表现出特质的性别差异信念。② 尽管对儿童关于特质的性别差异信念的研究结果不尽相同，但从已有研究中可以看出，至少在学前末期，儿童就已经能够对特质进行性别区分，或者说，儿童已经持有关于特质的性别差异信念。

值得注意的是，上述研究均在西方国家进行，其研究结果反映的是西方国家儿童性别差异信念的发展情况。相关研究表明，个体社会认知的发展存在文化差异③，儿童性别认知作为社会认知的重要内容，也必然存在文化差异。比如，一项跨文化研究发现，中国和以色列儿童在性别相关的推理和判断任务中表现出明显的差异，即与以色列儿童相比，中国儿童更能在特质等项目上进行性别区分。④ 鉴于此，有必要在中国文化背景下，探讨儿童关于特质的性别差异信念的发展特点。

性别图式理论认为，儿童一旦认同了自己的性别，就会选择适合自己性别的细节和脚本，表现出性别的内群体偏爱，并且对性别差异更加敏感。⑤ 从这个角度看，与异性特质相比，儿童对同性特

---

① Best D. L., Williams J. E., Cloud J. M., Davis S. W., Robertson L. S., Edwards J. R., Giles H., & Fowles J., "Development of Sex-Trait Stereotypes among Young Children in the United States, England, and Ireland", *Child Development*, Vol. 48, No. 4, August 1977, pp. 1375 – 1384.

② Ruble D., & Martin C., "Gender development", in W. Damon (Ed.), *Handbook of Child Psychology* (5th ed.), New York: Wiley, 1998, pp. 933 – 1016.

③ Han S., & Northoff G., "Culture-sensitive neural substrates of human cognition: a transcultural neuroimaging approach", *Nature Reviews Neuroscience*, Vol. 9, No. 8, February 2008, pp. 646 – 654.

④ Lobel T. E., Gruber R., Govrin N., & Mashraki-Pedhatzur, S., "Children's gender-related inferences and judgments: A cross-cultural study", *Developmental Psychology*, Vol. 37, No. 6, July 2001, pp. 839 – 846.

⑤ Martin C. L., & Halverson C. F., "A schematic processing model of sex typing and stereotyping in children", *Child Development*, No. 52, August 1981, pp. 1119 – 1134.

质的性别差异信念可能更强一些。另外，儿童早在 2 岁左右就形成了性别刻板印象，[①] 其性别刻板知识在学前期不断增长，并在 7 岁之后稳定发展。[②] 有研究者认为，性别差异信念与性别刻板印象密切相关。据此可以推测，关于特质的性别差异信念在学前阶段可能随年龄增长而逐渐增强，并在 7 岁左右处于稳定水平。因此，我们选取 3—8 岁儿童作为被试，在中国背景下考察儿童关于特质的性别差异信念及其发展特点。

二 方法

（一）被试

从 J 市三所幼儿园和一所小学选取 379 名儿童作为被试，其中，3 岁组 56 人，4 岁组 61 人，5 岁组 59 人，6 岁组 78 人，7 岁组 65 人，8 岁组 60 人，男女基本各半，平均年龄分别为 3.61 岁、4.44 岁、5.37 岁、6.50 岁、7.44 岁和 8.58 岁。

（二）研究材料

参考以往研究所使用的实验材料[③④]，初步选定 17 个特质项目，男性化和女性化项目基本各半。

选取 J 市一所普通幼儿园 4 岁、5 岁儿童 22 名（男女基本各

---

① Martin C. L., Ruble D. N., & Szkrybalo J., "Cognitive theories of early gender development", *Psychological Bulletin*, Vol. 128, No. 6, June 2002, pp. 903－933.

② 曹仁艳：《儿童性别刻板印象的发展与性别恒常性的关系：母亲教养态度的调节》，硕士学位论文，山东师范大学，2010 年。

③ 战欣：《儿童性别刻板印象的发展及其对社会判断的影响》，硕士学位论文，山东师范大学，2006 年。

④ Ruble D. N., Taylor L. J., Cyphers L., Greulich F. K., Lurye L. E., & Shrout P. E., "The role of gender constancy in early gender development", *Child Development*, Vol. 78, No. 4, May 2007, pp. 1121－1136.

半，平均年龄 5.1 ± 0.280 岁，预测被试均不参加正式施测）作为被试，问被试"谁经常是……（如，强壮的），男孩还是女孩"。根据被试所做回答，在男性和女性维度中各选取前 5 个最能代表男女性特质的特质词（共 10 项）作为正式施测项目。

（三）研究程序和方法

由经过培训的发展心理学专业的研究生担任主试，在幼儿园的会议室进行个别施测。首先告诉被试："一会儿我会给你说几个描述小朋友的词语，然后我会问你几个问题，你按照你的想法回答，好吗？"确定被试听懂后随机呈现各特质词，并向被试解释各特质词，比如，"强壮的"就是可以提起很重的东西；"爱帮助人的"就是经常帮助别的小朋友等，并问被试："谁经常这样，男孩还是女孩？"问题中，"男孩"和"女孩"的顺序随机呈现，主试随时记录被试的答案。

（四）计分方法

被试回答"男孩"计 0 分，回答"女孩"计 1 分。将男性化、女性化特质项目的得分平均后分别产生男性化特质性别分和女性化特质性别分。性别分范围为 0—1，0.5 为中性性别分。分值低于 0.5 表示某种特质与男孩联系更密切，分值高于 0.5 表示某种特质与女孩联系更密切。

三　结果

（一）3—8 岁儿童关于男性化特质的性别差异信念

为考察 3—8 岁儿童是否能将男性化特质与男孩联系，对其男性化特质性别分是否显著低于中性性别分（性别分 = 0.5）进行单

侧 t 检验。结果发现（见表 2 - 1），从总体及各年龄组来看，儿童的男性化特质性别分均显著低于中性性别分。这表明，各年龄组儿童能显著地把男性化特质与男孩相联系。

表 2 - 1　　3—8 岁儿童的男性化、女性化特质平均性别分及单侧 t 检验结果

	n	男性化特质		女性化特质	
		$\bar{x} \pm s$	t 值	$\bar{x} \pm s$	t 值
3 岁组	56	0.41 ± 0.23	- 2.975 **	0.62 ± 0.26	3.525 **
4 岁组	61	0.23 ± 0.25	- 8.454 ***	0.52 ± 0.27	0.608
5 岁组	59	0.14 ± 0.17	- 16.196 ***	0.63 ± 0.20	5.043 ***
6 岁组	78	0.06 ± 0.11	- 35.237 ***	0.86 ± 0.17	17.994 ***
7 岁组	65	0.02 ± 0.06	- 66.543 ***	0.94 ± 0.12	30.748 ***
8 岁组	60	0.04 ± 0.08	- 44.198 ***	0.97 ± 0.08	48.092 ***
总体	379	0.14 ± 0.21	- 33.430 ***	0.77 ± 0.26	20.072 ***

注:* 表明与中性性别分存在差异的程度, ** 表示 P < 0.01, *** 表示 P < 0.001。

为考察 3—8 岁儿童关于男性化特质的性别差异信念的性别与年龄特点，对男性化特质性别分进行 2（性别）×6（年龄组）的方差分析。结果发现，性别主效应显著，$F(1, 367) = 5.242$，P < 0.05，男孩的男性化特质性别分显著低于女孩的男性化特质性别分，这说明，与女孩相比，男孩更倾向于将男性化特质与男孩相联系，即他们有关男性化特质的性别差异信念较女孩更强一些；年龄组主效应显著，$F(5, 367) = 47.308$，P < 0.001。进一步事后检验（LSD）发现，3 岁组儿童男性化特质性别分显著高于后 5 个年龄组，4 岁组儿童性别分显著高于后 4 个年龄组，5 岁组儿童性别分显著高于后 3 个年龄组，后 3 个年龄组儿童性别分不存在显著差异。总体来看，儿童关于男性化特质的性别差异信念随年龄增长不断增强，6 岁之后趋于稳定。性别与年龄组之间的交互作用（见

图 2 - 1）显著，F（5，367）＝2.651，P＜0.05。简单效应分析发现，3 岁组儿童的男性化特质性别分存在显著的性别差异，t ＝ - 2.098，P＜0.05，3 岁组男孩的男性化特质性别分显著低于 3 岁组女孩的男性化特质性别分；4 岁组儿童的男性化特质性别分的性别差异边缘显著，t ＝ - 1.698，P＝0.095，4 岁组男孩的男性化特质性别分边缘显著地低于 4 岁组女孩的男性化特质性别分；5—8 岁组儿童的男性化特质性别分性别差异均不显著，t 值分别为 - 0.988、0.813、- 0.039、0.478，ps＞0.05。这表明，3 岁组和 4 岁组男孩关于男性化特质的性别差异信念均强于女孩，而其余各年龄组儿童均不存在这样的特点。

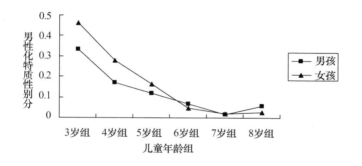

图 2 - 1　各年龄组男孩和女孩的男性化特质性别分

（二）3—8 岁儿童关于女性化特质的性别差异信念

为考察 3—8 岁儿童是否能将女性化特质与女孩相联系，对其女性化特质性别分是否显著高于中性性别分进行单侧 t 检验。结果发现（见表 2 - 1），从总体上看，除 4 岁组外，各个年龄组儿童的女性化特质性别分均显著高于中性性别分，即他们能显著地把女性化特质与女孩联系。

为考察 3—8 岁儿童关于女性化特质的性别差异信念的性别与

年龄特点，对女性化特质的性别分进行 2（性别）×6（年龄组）的方差分析。结果发现，性别主效应显著，$F(1, 367) = 13.649$，P<0.001，男孩的女性化特质性别分显著低于女孩的女性化特质性别分，这表明，与男孩相比，女孩更倾向于将女性化特质与女孩相联系，即她们有关女性化特质的性别差异信念较男孩更强一些；年龄组主效应显著，$F(5, 367) = 58.983$，P<0.001。事后检验（LSD）发现，3 岁组儿童性别分与 5 岁组之间无显著差异，但显著高于 4 岁组，显著低于 6 岁、7 岁、8 岁组，4 岁组性别分显著低于3 岁组和后 4 个年龄组，5 岁组性别分显著低于后 3 个年龄组，6 岁组性别分显著低于 7 岁、8 岁组，7 岁、8 岁组性别分之间无显著差异。也就是说，除 4 岁儿童外，其余各年龄组儿童的女性化特质性别分逐渐升高，即关于女性化特质的性别差异信念逐渐增强，6 岁后处于相对稳定水平。性别与年龄组之间交互作用（见图 2 - 2）不显著，$F(5, 367) = 1.022$，P>0.05。

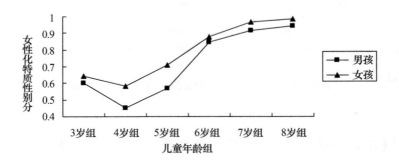

图 2 - 2　各年龄组男孩和女孩的女性化特质性别分

四　讨论

本研究发现，从总体上看，3—8 岁儿童均倾向于将男性化特质与男孩相联系，将女性化特质与女孩相联系，即 3—8 岁儿童均持有关于特质的性别差异信念。从各年龄组来看，3—5 岁儿童越来越倾向于将男性化特质与男孩相联系，将女性化特质与女孩相联系，6 岁之后这种倾向趋于稳定。换言之，在学前阶段，儿童关于特质的性别差异信念随着年龄增长变得越来越强，并在 6 岁左右达到稳定水平。这可能是由于：一方面，随着年龄的增长，儿童越来越多地观察到特质相关行为上的性别差异。比如，儿童在游戏中认识到女孩总是爱哭，并且胆小，而男孩比较勇敢，喜欢打架。另一方面，根据模糊—痕迹理论，个体对行为信息的编码逐渐从完整、准确的表征发展为模糊、简要的表征（如特质），并且随着年龄的增长，个体越来越偏爱利用模糊痕迹表征行为。[①] 从上述两点可以看出，儿童关于特质的性别差异信念随着年龄的发展逐渐增强。而在 6 岁左右，儿童关于特质的性别差异信念已经达到较为成熟水平，因此，6—8 岁儿童关于特质的性别差异信念没有表现出年龄差异。但是，在本研究中，4 岁儿童不具有关于女性化特质的性别差异信念，其原因有待进一步探讨。

在性别差异方面，从总体上看，男孩比女孩更多地将男性化特质归于男孩，女孩比男孩更多地将女性化特质归于女孩。换言之，儿童对同性特质的性别差异信念强于对异性特质的性别差异信念。其原因可能在于：依照性别图式理论，儿童总是非常积极主动地寻找并获得与他们自身性别相一致的兴趣、价值观、行为方式和特质等信息。[②③] 因此，与异性特质相比，儿童对同性特质的认识更多一些。同时，有关儿童同伴关系的研究表明，儿童在学前阶段和学

---

① Reyna V. F. , & Brainerd C. J. , "Fuzzy-trace theory: An interim synthesis", *Learning and Individual Differences* , Vol. 7 , No. 1 , May 1995 , pp. 1 - 75.

② Martin C. L. , & Halverson C. F. , "A schematic processing model of sex typing and stereotyping in children", *Child Development* , No. 52 , June 1981 , pp. 1119 - 1134.

③ Serbin L. A. , Polishta K. K. , & Gulko J. , "The development of sex-typing in middle school", *Monographs of the Society for Research in Child Development* , Vol. 58 , No. 2 , March 1993 , pp. 1 - 99.

龄初期更喜欢与同性儿童交往①②,在这样的同伴交往中,儿童可以认识到更多同性的特质,所以对同性特质的了解更多一些。基于对同性特质较多的了解,儿童认为,同性比异性更多地表现这些特质,儿童对这些特质也更能进行性别区分,因此对同性特质的性别差异信念强于对异性特质的性别差异信念。从各年龄组看,与同龄女孩相比,3 岁和 4 岁男孩均倾向于将男性化特质归于男孩,而其余各年龄组儿童并没有表现出这种性别差异。其原因可能是:在本研究中,有 96.9% 的 5 岁男孩和 92.3% 的 5 岁女孩均将男性化特质归于男孩。也就是说,在儿童 5 岁时,他们关于男性化特质的性别差异信念已经发展良好,男女儿童在特质上进行性别区分的能力已经达到较高水平,因此并未表现出性别差异。

综合本研究结果,从整体上看,儿童从 3 岁起便持有关于特质的性别差异信念,这些信念随年龄增长逐渐增强,6 岁之后稳定发展;并且,儿童关于同性特质的性别差异信念强于对异性特质的性别差异信念。值得注意的是,本研究仅仅是通过外显的方法考察儿童关于特质的性别差异信念。而有研究发现,性别差异信念的内隐测量并不明显地提到性别,可以在一定程度上减少社会期望对研究结果的影响。③ 因此,未来的研究可以采用内隐测量的方法进一步考察儿童关于特质的性别差异信念的特点。

---

① Serbin L. A., Polishta K. K., & Gulko J., "The development of sex-typing in middle school", *Monographs of the Society for Research in Child Development*, Vol. 58, No. 2, March 1993, pp. 1 – 99.

② Sippola L. K., Bukowski W. M., & Noll R. B., "Dimensions of liking and disliking underlying the same-sex preference in childhood ad early adolescence", *Merrill-Palmer Quarterly*, No. 43, February 1997, pp. 591 – 609.

③ Hudley C., & Graham S., "Stereotypes of achievement striving among early adolescents", *Social Psychology of Education*, Vol. 5, No. 2, May 2001, pp. 201 –224.

第 三 章

反刻板印象、内隐性别刻板印象与情绪

第一节　反刻板印象与内隐性别刻板印象的关系

反刻板印象（counterstereotype），是指符合刻板印象的群体在某一方面的特征表现得与刻板印象相反，违背了人们对这些特征的期望和要求的现象。反刻板印象主要在一些方面体现：在人格特征上，人们认为女性是"依赖他人的""胆小的"，认为男性是"独立的""勇敢的"，如果女性表现出独立、勇敢，那么便会被认为是反刻板印象的代表[①]；在能力表现上，人们认为男性在各方面更优秀，而女性则稍弱一些，那么，如果女性表现出具有较高的才能，而男性更加依赖他人，也会成为反刻板印象的典型；在家务分工上，中国传统的观点认为"男主外、女主内"，如果一个男人整日操持家务，而女性在事业上拼搏，也会被认为是反刻板印象的代表。另外，在职业选择以及学科的选择上，都有反刻板印象的典型。可以说，刻板印象涉及的领域，都会有反刻板印象存在。那么，能否说个体持有的反刻板印象与刻板印象是一

---

① 秦启文、余华：《性别角色刻板印象的调查》，《心理科学》2001 年第 5 期。

一对应的呢？

Stangor 等提出的语义网络模型认为，刻板印象是由社会类别（category）与特质概念（traits）相互联结组成的"语义网络系统"。[①] 当刻板印象被激活时，一些在语义上相互联系的概念也会被激活。随后，Dijksterhuis 和 Khippenberg 对语义网络模型进行了扩展，他们指出，在语义网络系统中，刻板印象除了与刻板一致概念之间存在正向联结，还存在刻板印象和与其不一致的特质概念之间的负向联结。[②] 因此可以看出，反刻板印象与刻板印象是一一对应的，并且，刻板印象与一致概念和不一致概念之间的联结是同时存在的。那么，如果刻板印象与不一致概念之间的联结强度大于刻板印象与一致概念之间的联结强度时，或者说，当反刻板联结的强度比刻板化联结强度更大时，刻板印象就会被削弱，甚至消除。

许多研究者致力于削弱或消除刻板印象，方法包括在前意识阶段阻止刻板印象的激活，以及刻板印象产生后有意地控制它对行为的影响。[③] Bodenhausen 和 Macrae 对上述两种方法进行比较发现，较后者而言，第一时间阻止刻板印象激活对于削弱甚至消除内隐刻板印象更为有效。[④]

---

① Stangor C. , & Ford T. E. , "Accuracy and expectancy-confirming orientations and the development of stereotypes and prejudice", *European Review of Social Psychology*, New York：Wiley, No. 3, 1992, pp. 57 – 89.

② Dijksterhuis A. , & van Knippenberg A. , "The knife that cuts both ways：Facilitated and inhibited access to traits as a result of stereotype-activation", *Journal of Experimental Social Psychology*, No. 32, March 1996, pp. 271 – 288.

③ 庞晓佳、张大均、王鑫强、王金良：《刻板印象干预策略研究述评》，《心理科学进展》2011 年第 2 期。

④ Bodenhausen G. V. , & Macrae C. N. , "Stereotype activation and inhibition", in R. S. Wyer Jr. （Ed. ）, *Advances in Social Cognition*, Mahwah, NJ：Erlbaum, Vol. 11, 1998, pp. 1 – 52.

　　抑制刻板激活的一个方法是非刻板联结的训练，这种训练包括否定刻板印象（negating stereotypes）和确认反刻板印象（affirming counterstereotypes）。① 比如，在 Kawakami, Dovidio, Moll, Hermsen 和 Russin 的实验中，首先采用 Stroop 范式考察被试的刻板激活，接着对被试进行训练。② 在训练中，向被试呈现黑人和白人的照片以及与黑人和白人刻板有关的特质词，被试的任务是当刻板一致的照片—特质词配对出现时按"NO"键反应（如，黑人照片—黑人刻板化特质词），当刻板不一致的照片—特质词配对出现时按"YES"键反应（如，黑人照片—白人刻板化特质词）。训练之后再完成一项考察刻板激活的 Stroop 任务。结果发现，在接受训练之后，被试的刻板激活被显著地抑制。Kawakami 等人认为，刻板激活被减弱的机制可能是对刻板一致信息重复反应"NO"会破坏记忆中的刻板联结，也可能是对刻板不一致信息重复反应"YES"使被试获得新的反刻板联结。③ Gawronski 等人对这两种机制进行了验证。他们向被试呈现男性与女性的名字以及男性化、女性化特质词。其中，对一半被试进行的训练是"否定刻板印象"，即要求他们只对刻板化配对按"NO"键反应（如，男

　　① Gawronski B. , Deutsch R. , Mbirkou S. , Seibt B. , & Strack F. , "When 'just say no' is not enough: Affirmation versus negation training and the reduction of automatic stereotypes activation", *Journal of Experimental Social Psychology*, No. 44, October 2008, pp. 370 – 377.

　　② Kawakami K. , Dovidio J. F. , Moll J. , Hermsen S. , & Russin A. , "Just say no (to stereotyping): effects of training in the negation of stereotypic associations on stereotypic activation", *Journal of Personality and Social Psychology*, No. 78, August 2000, pp. 871 – 888.

　　③ Kawakami K. , Dovidio J. F. , Moll J. , Hermsen S. , & Russin A. , "Just say no (to stereotyping): effects of training in the negation of stereotypic associations on stereotypic activation", *Journal of Personality and Social Psychology*, No. 78, August 2000, pp. 871 – 888.

性姓名与"strongness"的配对，女性姓名与"weakness"的配对），
对反刻板化配对不做反应；而对另一半被试进行的训练是"确认反
刻板印象"，即只有当反刻板化配对出现时按"YES"键反应（如，
男性姓名与"weakness"的配对，女性姓名与"strong"的配对），
而对刻板化配对不做反应。[①] 结果表明，只有"确认反刻板印象"
的训练能显著抑制被试的刻板激活，而"否定刻板印象"的训练
不但没有抑制刻板激活，反而具有一定的促进作用。换言之，通
过确认反刻板印象样例来启动反刻板能有效抑制自动化的刻板激
活。许多研究证实了这一观点。比如，在 Blair 等人的研究中，一
部分被试在开始实验前要想象一位女性领导人的形象。[②] 结果发
现，与其他被试相比，想象女性领导人形象的被试表现出了更少
的性别刻板印象。

第二节　情绪对刻板印象的影响

一　情绪对社会认知的影响

在过去的几十年间，许多研究者致力于个体的情绪状态对
其认知加工的影响，包括情绪影响个体的感知、推理以及判

---

① Gawronski B., Deutsch R., Mbirkou S., Seibt B., & Strack F., "When'just say no'is not enough: Affirmation versus negation training and the reduction of automatic stereotypes activation", *Journal of Experimental Social Psychology*, No. 44, February 2008, pp. 370 – 377.

② Blair I. V., Ma J. E., & Lenton A. P., "Imaging stereotypes away: The moderation of implicit stereotypes through mental imagery", *Journal of Personality and Social Psychology*, No. 81, February 2001, pp. 828 – 841.

断。①②③④⑤ 那么，情绪是怎样对个体的认知加工产生影响的呢？

（一）情绪影响认知的相关理论

1. 情绪—信息加工模型（The Affect-and-information-processing Model）

根据情绪—信息加工模型，个体的情绪状态能够影响其信息加工。⑥⑦⑧⑨ 具体来说，悲伤的情绪促进个体精细的（或系统的）

---

①　Forgas J. P., & Fiedler K., "Us and them: Mood effects on inter-group discrimination", *Journal of Personality and Social Psychology*, No. 70, March 1996, pp. 28 – 40.

②　Isen A. M., "Toward understanding the role of affect in cognition", in R. S. Wyer & T. K. Krull (Eds.), *Handbook of Social Cognition*, Hillsdale, NJ: Lawrence Erlbaum, Vol. 3, 1984, pp. 179 – 236.

③　Isen A. M., "Positive affect, cognitive processes, and social behaviour", in L. Berkowitz (Ed.), *Advances in Experimental Social Psychology*, New York: Academic Press, Vol. 20, 1987, pp. 203 – 253.

④　Mackie D. M., & Hamilton D. L. (Eds.), *Affect, Cognition, and Stereotyping: Interactive Processes in Group Perception*, San Diego, CA: Academic Press, 1993.

⑤　Sinclair R. C., & Mark M. M., "The influence of mood state on judgment and action: Effects on persuasion, categorization, social justice, person perception, and judgmental accuracy", in L. L. Martin & A. Tesser (Eds.), *The Construction of Social Judgments*, Hillsdale, NJ: Lawrence Erlbaum, 1992, pp. 165 – 193.

⑥　Bless H., Bohner G., Schwarz N., & Strack F., "Mood and persuasion: A cognitive response analysis", *Personality and Social Psychology Bulletin*, No. 16, June 1990, pp. 331 – 345.

⑦　Bodenhausen G. V., "Emotions, arousal, and stereotypic judgments: A heuristic model of affect and stereotyping", in D. M. Mackie & D. L. Hamilton (Eds.), *Affect, Cognition, and Stereotyping: Interactive Processes in Group Perception*, San Diego, CA: Academic Press, 1993, pp. 13 – 37.

⑧　Mackie D. M., & Worth L. T., "Processing deficits and the mediation of positive affect in persuasion", *Journal of Personality and Social Psychology*, No. 57, August 1989, pp. 27 – 40.

⑨　Schwarz N., Bless H., & Bohner G., "Mood and persuasion: Affective states influence the processing of persuasive communications", in M. P. Zanna (Ed.), *Advances in Experimental Social Psychology*, New York: Academic Press, Vol. 24, 1991, pp. 161 – 197.

加工[①]，而高兴的情绪抑制精细的信息加工，并导致个体更依赖判断中表面的启发式线索。Bodenhausen 等人认为，刻板印象可以被看作当个体缺乏某种能力或动机去思考对目标个体的归因时所依赖的启发式线索。[②③④⑤] 也就是说，根据这个理论，消极情绪（如悲伤）促进个体进行精细的信息加工，而积极情绪（如高兴）促进个体的自动化信息加工。许多研究的结果也与此观点一致。比如，有研究发现，在社会判断中，高兴的被试更依赖类别化信息（如刻板印象）[⑥]，而悲伤的被试更依赖于个体化信息[⑦]。

---

① Chaiken S. , Liberman A. , & Eagly A. H. , "Heuristic and systematic information processing within and beyond the persuasion context", in J. S. Uleman & J. A. Bargh (Eds.), *Unintended Thought*, New York: Guilford, 1989, pp. 212 – 252.

② Bodenhausen G. V. , "Stereotypes as judgmental heuristics: Evidence of circadian variations in discrimination", *Psychological Science*, No. 1, October 1990, pp. 319 – 322.

③ Bodenhausen G. V. , "Emotions, arousal, and stereotypic judgments: A heuristic model of affect and stereotyping", in D. M. Mackie & D. L. Hamilton (Eds.), *Affect, Cognition, and Stereotyping: Interactive Processes in Group Perception*, San Diego, CA: Academic Press, 1993, pp. 13 – 37.

④ Bodenhausen G. V. , Sheppard L. A. , & Kramer G. P. , "Negative affect and social judgment: The differential impact of anger and sadness", *European Journal of Social Psychology*, No. 24, April 1994, pp. 45 – 62.

⑤ Bodenhausen G. V. , & Wyer R. S. , "Effects of stereotypes on decision making and information-processing strategies", *Journal of Personality and Social Psychology*, No. 48, December 1985, pp. 267 – 282.

⑥ Bodenhausen G. V. , Kramer G. P. , & Sasser K. , "Happiness and stereotypic thinking in social judgment", *Journal of Personality & Social Psychology*, Vol. 66, No. 4, June 1994, pp. 621 – 632.

⑦ Bless H. , Schwarz N. , & Wieland R. , "Mood and the impact of category membership and individuating information", *European Journal of Social Psychology*, No. 26, March 1996, pp. 935 – 959.

许多假说对这个模型作出了解释。简化加工假说（The Re-duced-processing Hypothesis）认为，"高兴"能减弱人们加工进入大脑的信息的能力，因此个体只能对简单的信息（如，个体已经掌握的知识、经验）进行加工。认知假说（The Cognitive Hypothesis）①②③④认为，"高兴"降低了个体认知资源的总量⑤，因此，这种情绪下的个体没有更多的认知资源投入其他的任务中，只能依赖不太需要认知努力的自动化加工。享乐主义假说（The He-donic View）认为，一方面，高兴的个体为了维持他们的积极情绪，将他们的认知资源分配给能够维持积极情绪的任务⑥，因此，他们不会在其他任务中投入过多认知资源，除非那种任务有利于维持当前的积极情绪⑦；另一方面，悲伤的个体要投入更多的努

① Asuncion A. G., & Lam W. F., "Affect and impression formation: Influence of mood on person memory", *Journal of Experimental Social Psychology*, No. 31, July 1995, pp. 437 – 464.

② Isen A. M., "Positive affect, cognitive processes, and social behaviour", in L. Berkowitz (Ed.), *Advances in Experimental Social Psychology*, New York: Academic Press, Vol. 20, 1987, pp. 203 – 253.

③ Mackie D. M., & Worth L. T., "Processing deficits and the mediation of positive affect in persuasion", *Journal of Personality and Social Psychology*, No. 57, October 1989, pp. 27 – 40.

④ Worth L. T., & Mackie D. M., "Cognitive mediation of positive affect in persua-sion", *Social Cognition*, No. 5, March 1987, pp. 76 – 94.

⑤ Ellis H. C., & Ashbrook P. W., "Resource allocation model of the effects of de-pressed mood states on memory", in K. Fiedler & J. Forgas (Eds.), *Affect, Cognition, and Social Behaviour*, Toronto: Hogrefe, 1988, pp. 25 – 43.

⑥ Isen A. M., "Positive affect, cognitive processes, and social behaviour", in L. Berkowitz (Ed.), *Advances in Experimental Social Psychology*, New York: Academic Press, Vol. 20, 1987, pp. 203 – 253.

⑦ Bodenhausen G. V., Kramer G. P., & Sasser, K., "Happiness and stereotypic thinking in social judgment", *Journal of Personality & Social Psychology*, Vol. 66, No. 4, April 1994, pp. 621 – 632.

力去加工信息，以改变当前消极不良的情绪状态①②。机能主义假说（The Functionalist Approach）认为，情绪状态对个体具有重要功能：它意味着个体当前所处环境的好坏。③④ 根据这个假说，消极的情绪状态（如，悲伤）代表着个体当前所处的环境是有问题的，并且它代表着"这个环境缺乏积极的结果或者被消极的结果所威胁"，这就会促使个体改变当前环境。要改变当前环境，个体必须通过对进入个体的信息进行精细的加工来对环境进行准确的表征。相反，积极情绪（如，高兴）意味着当前的环境没有问题，"是安全、并令人满意的"。⑤ 在这种环境中，个体不必强迫自己去付出认知努力，他们更倾向于依赖启发式线索（如，刻板印象）。

综上所述，根据情绪—信息加工模型，当个体要对目标做出判断时，消极情绪的个体可能更依赖于个体化信息，而忽略目标所属群体的刻板印象。与此相反，积极情绪的个体更依赖于刻板印象，并且很少注意到环境提供的个体信息，或者至少不愿意为加工不一

---

① Lassiter G. D., Koenig L. J., & Apple K. J., "Mood and behavior perception: Dysphoria can increase and decrease effortful processing of information", *Personality and Social Psychology Bulletin*, No. 22, June 1996, pp. 794 – 810.

② Wegener D. T., Petty R. E., & Smith S. M., "Positive mood can increase or decrease message scrutiny: The hedonic contingency view of mood and message processing", *Journal of Personality and Social Psychology*, No. 69, September 1995, pp. 5 – 15.

③ Frijda N., "The laws of emotion", *American Psychologist*, No. 43, December 1988, pp. 349 – 358.

④ Schwarz N., "Feeling as information: Informational and motivational functions of affective states", in E. T. Higgins & R. M. Sorrentino (Eds.), *Handbook of Motivation and Cognition*, New York: Guilford, Vol. 2, 1990, pp. 527 – 561.

⑤ Schwarz N., "Feeling as information: Informational and motivational functions of affective states", in E. T. Higgins & R. M. Sorrentino (Eds.), *Handbook of Motivation and Cognition*, New York: Guilford, Vol. 2, 1990, pp. 527 – 561.

致信息付出认知努力。[①②]

2. 情绪一般知识模型（The Mood-and-General-Knowledge Model，MAGK）

Bless 等人在机能主义假说的基础上提出了情绪一般知识模型。[③④⑤] 尽管这个模型与机能主义假说存在一定的相似之处，但它们在一个重要方面有所不同。与前面提到的模型和假说相比，MAGK 模型认为，情绪对认知的影响不再被看作加工能力或动机的变化。[⑥⑦⑧] 一些研究的结果也支持了 MAGK 模型这一方面的观点，

①　Bless H. , Schwarz N. , & Wieland R. , "Mood and the impact of category membership and individuating information", *European Journal of Social Psychology*, No. 26, June 1996, pp. 935 – 959.

②　Bodenhausen G. V. , Kramer G. P. , & Sasser K. , "Happiness and stereotypic thinking in social judgment", *Journal of Personality & Social Psychology*, Vol. 66, No. 4, July 1994, pp. 621 – 632.

③　Bless H. , Clore G. L. , Schwarz N. , Golisano V. , Rabe C. , & Wölk M. , "Mood and use of scripts: Does a happy mood really lead to mindlessness?" *Journal of Personality and Social Psychology*, No. 71, August 1996, pp. 665 – 679.

④　Bless H. , Schwarz N. , & Kemmelmeier M. , "Mood and stereotyping: Affective states and the use of general knowledge structures", in W. Stroebe & M. Hewstone (Eds.), *European Review of Social Psychology*, Chichester, UK: Wiley, Vol. 7, 1996, pp. 63 – 93.

⑤　Bless H. , Schwarz N. , & Wieland R. , "Mood and the impact of category membership and individuating information", *European Journal of Social Psychology*, No. 26, April 1996, pp. 935 – 959.

⑥　Bless H. , Clore G. L. , Schwarz N. , Golisano V. , Rabe C. , & Wölk M. , "Mood and use of scripts: Does a happy mood really lead to mindlessness?" *Journal of Personality and Social Psychology*, No. 71, December 1996, pp. 665 – 679.

⑦　Levine L. J. , & Burgess S. L. , "Beyond general arousal: Effects of specific emotions on memory", *Social Cognition*, No. 15, November 1997, pp. 157 – 181.

⑧　Bless H. , Schwarz N. , & Kemmelmeier M. , "Mood and stereotyping: Affective states and the use of general knowledge structures", in W. Stroebe & M. Hewstone (Eds.), *European Review of Social Psychology*, Chichester, UK: Wiley, Vol. 7, 1996, pp. 63 – 93.

比如，有研究发现，与消极情绪的被试相比，积极情绪的被试在创造力或问题解决任务中表现得更好①，并且能够进行更为精密的信息加工。② 根据 MAGK 模型，积极情绪个体表现出的简化的信息加工不再是因为加工动机或者能力的降低，而是由于他们对一般知识结构（如图式、刻板印象）的信任增强。Bless 等人也认为，积极情绪的个体依赖一般知识结构并非简化加工的结果，而是其原因。③

与机能主义假说一致的是，MAGK 模型认为，情绪状态代表人们所处环境的好坏，并且，情绪状态也能告知个体在多大程度上依赖一般知识结构。消极情绪（如，悲伤）意味着当前的处境存在问题，在这样的环境中，此时依赖一般知识结构可能会使个体感到适应不良、不安全。因此，消极情绪的个体倾向于注重任务中的具体信息以帮助其自身解决环境中的问题。④ 相反，积极情绪（如，高兴）意味着环境是安全可靠的，在这样的环境中，简单地依赖一般知识结构是合适的，因为这样可以使个体节省认知资源以投入其他任务中。⑤ 由于积极情绪的个体十分信任他们现有的知识结构，甚

---

① Isen A. M., "Positive affect, cognitive processes, and social behaviour", in L. Berkowitz (Ed.), *Advances in Experimental Social Psychology*, New York: Academic Press, Vol. 20, 1987, pp. 203 – 253.

② Martin C. L., Ward D. W., Achee J., & Wyer R. S., "Mood as input: People have to interpret the motivational implications of their moods", *Journal of Personality and Social Psychology*, No. 64, June 1993, pp. 317 – 326.

③ Bless H., Schwarz N., & Kemmelmeier M., "Mood and stereotyping: Affective states and the use of general knowledge structures", in W. Stroebe & M. Hewstone (Eds.), *European Review of Social Psychology*, Chichester, UK: Wiley, Vol. 7, 1996, pp. 63 – 93.

④ Lambert A. J., Khan S. R., Lickel B. A., & Fricke K., "Mood and the correction of positive versus negative stereotypes", *Journal of Personality and Social Psychology*, No. 72, October 1997, pp. 1002 – 1016.

⑤ Bless H., Clore G. L., Schwarz N., Golisano V., Rabe C., & Wölk M., "Mood and use of scripts: Does a happy mood really lead to mindlessness?" *Journal of Personality and Social Psychology*, No. 71, August 1996, pp. 665 – 679.

至努力将环境所提供的具体信息与其知识结构相联系。个体的一般知识结构与环境提供的具体信息之间不一致的程度会决定积极情绪的个体采用何种策略来处理这种不一致的关系。具体来说，当个体的一般知识结构与环境中的具体信息之间存在轻微的不一致时，个体为了使具体信息与其一般知识结构相一致而采用"再评价策略"（reevaluation strategy）。这样，积极情绪的个体只是依赖其一般知识结构进行信息加工。然而，随着二者之间不一致的程度越来越大，积极情绪的个体往往会采用两种策略：首先，他们可能会通过对目标进行重新分类以整合不一致信息。①② 其次，当二者不一致程度最大化，即个体的一般知识结构与环境提供的具体信息完全相反时，个体不能对目标进行重新分类，他们会放弃一般知识结构而仅仅依赖环境提供的数据或者目标的具体信息。③④

（二）社会认知研究中情绪诱发的方法

在情绪影响认知的研究中，情绪诱发是最为重要的内容，它不仅关系到研究者能否获得想要的情绪状态，也关系到所诱发的情绪是否稳定、有效，并且与实验研究的成功与否息息相关。情绪诱发

---

① Fiske S. T., & Neuberg S. L., "A continuum of impression formation from category-based to individuating processes: Influences of information and motivation on attention and interpretation", in M. P. Zanna (Ed.), *Advances in Experimental Social Psychology*, San Diego, CA: Academic Press, Vol. 23, 1990, pp. 1 – 74.

② Hewstone M., "Revision and change of stereotypic beliefs: In search for the elusive subtyping model", in W. Stroebe & M. Hewstone (Eds.), *European Review of Social Psychology*, Chichester, UK: Wiley, Vol. 5, 1994, pp. 69 – 108.

③ Fiske S. T., & Neuberg S. L., "A continuum of impression formation from category-based to individuating processes: Influences of information and motivation on attention and interpretation", in M. P. Zanna (Ed.), *Advances in Experimental Social Psychology*, San Diego, CA: Academic Press, Vol. 23, 1990, pp. 1 – 74.

④ Srull T. K., & Wyer R. S., "Person memory and judgment", *Psychological Review*, No. 96, May 1989, pp. 58 – 83.

的方法，是指在非自然和严格控制的条件下唤起个体临时性情绪状态的策略。[①] 在过去几十年的研究中，研究者设计出一系列诱发情绪的方法，主要包括自传式回忆法、想象情绪诱发、音乐情绪诱发、视频情绪诱发等。

1. 自传式回忆法

在阿德勒心理疗法中，自传式回忆是一种重要的心理咨询手段，它被广泛地应用于心理诊断与咨询的临床实践中。后来，Brewer 和 Doughtie 将其进行改进，从而用于情绪的相关研究。[②] 在诱发被试的情绪时，首先要求被试闭上眼睛，回忆 3 件能分别唤起悲伤、愉快等情绪的事件，随着回忆的进行，被试会越来越能感受到悲伤、愉快等情绪。有研究者认为，虽然这种方法能够有效诱发出想要的情绪，但仍存在一定的缺陷，比如，这种方法要求被试有意识的合作，这就可能会导致要求特征的出现。[③]

2. 想象情绪诱发

想象情绪诱发方法最初由 Wright 和 Mischel 提出。这种方法要求被试首先在主试的指导语帮助下放松全身，随后，被试要集中精力注意主试的指导语，按照要求想象一些场景。这些场景既可以是真实发生过的，也可以是被试自己想象的，但都要栩栩如生。被试

---

① Banos R., Liano V., & Botella C., Changing Induced Moods Via Virtual Reality, Jsselsteijn, W. I., Lecture Notes in Computer Science: Persuasive Technology. Heidelberg: Spriger Berlin, 2006, pp. 7 – 15.

② Brewer D., & Doughtie E. B., "Induction of Mood and Mood Shift", *Journal of Clinical Psychology*, Vol. 36, No. 1, March 1980, pp. 215 – 226.

③ Westermann R., Spies K., & Stahl G., "Relative Effectiveness and Validity of Mood Induction Procedures: A Meta-Analysis", *European Journal of Social Psychology*, Vol. 26, No. 4, June 1996, pp. 557 – 580.

要认真感受并思考这些场景。[①] 有的研究者为了使情绪诱发更加有效，还让被试把刚才想象的场景描述出来。这种方法与自传式回忆法相似，因此仍无法避免自传式回忆法的弊端。[②]

3. 音乐情绪诱发

这种方法是向被试播放具有强烈情绪色彩的音乐，唤起被试相应的情绪体验。这种音乐情绪诱发被广泛应用于消费者行为控制、情绪紊乱的心理治疗、个体自我情绪调节等领域。[③] 经过众多研究成果的不断积累，一些音乐已经与特定情绪形成稳定的对应关系，比如，巴赫的《勃兰登堡协奏曲》、贝多芬的《第六交响乐》能够诱发愉悦的情绪体验；霍尔斯特的《火星：战争使者》能够诱发恐惧的情绪体验；巴伯的《弦乐柔板》能诱发悲伤的情绪体验。[④⑤] 有研究者认为，使用音乐作为诱发情绪的手段，具有其他方法难以比拟的优势：第一，音乐能诱发一系列不同的情绪体验，如愉悦、悲伤、恐惧等；第二，与想象策略相比，音乐诱发的情绪体验往往

---

① Wright J., Mischel W., "Influence of affective on cognitive social learning person variables", *Journal of Personality and Social Psychology*, Vol. 43, No. 5, April 1982, pp. 901 – 914.

② Westermann R., Spies K., Stahl G., "Relative Effectiveness and Validity of Mood Induction Procedures: A Meta-Analysis", *European Journal of Social Psychology*, Vol. 26, No. 4, May 1996, pp. 557 – 580.

③ Gold C., Voracek M., Wigram T., "Effects of music therapy for children and adolescents with psychopathology: A meta-analysis", *Journal of Child Psychology and Psychiatry*, Vol. 45, No. 6, July 2004, pp. 1054 – 1063.

④ Peretz I., Gagnon L., Bouchard B., "Music and emotion: perception determinants, immediacy, and isolation after brain damage", *Cognition*, Vol. 68, No. 2, October 1998, pp. 111 – 141.

⑤ Baumgartner T., Esslen M., Jancke L., "From emotion perception to experience: Emotions evoked by pictures and classic music", *Inernational Journal of Psychophysiology*, Vol. 60, No. 1, September 2006, pp. 34 – 43.

更为深入、持久；第三，音乐诱发情绪法具有良好的跨文化一致性。[①] 但是，音乐诱发情绪的方法存在的最大问题在于标准化情绪诱发材料库的缺乏[②]，并且，有研究者认为，音乐所诱发的情绪，与日常生活中自然产生的情绪是否一致有待进一步探讨。[③]

4. 视频情绪诱发

视频情绪诱发，是指通过观看视频片段来诱发被试产生特定的情绪体验的方法。一般而言，相关领域的研究者经常剪辑某些特定的电影片段作为视频诱发情绪的材料。比如，有研究者采用电影《当哈利遇上莎莉》来诱发积极情绪，用《舐犊情深》来诱发消极情绪。[④] 与其他诱发情绪的手段相比，视频材料属于动态材料，集视、听于一体，因此囊括了视觉、听觉刺激材料的所有优点。此外，Gilet 认为，在道德问题方面，视频材料要明显优于其他材料，因此，视频情绪诱发比其他方法更为有效。[⑤]

（三）社会认知研究中情绪评定的方法

目前，情绪评定的方法大多采用自我报告法，常用的问卷有以下几种：（1）积极情感—消极情感量表（Positive Affectivity Nega-

---

① Fritz T., Jentschke S., Gosselin N., Sammler D., Peretz I., Turner R., et al., "Universal recognition of three basic emotions in music", *Current Biology*, No. 19, April 2009, pp. 573 – 576.

② 郑璞、刘聪慧、余国良:《情绪诱发方法述评》,《心理科学进展》2012 年第 1 期。

③ Konečni V. J., "Does music induce emotion? A theoretical and methodological analysis", *Psychology of Aesthetics*, *Creativity and the Arts*, No. 2, May 2008, pp. 115 – 129.

④ Gilet A. L., "Mood induction procedures: A critical review", *Encephale*, Vol. 34, No. 3, March 2008, pp. 233 – 239.

⑤ Gilet A. L., "Mood induction procedures: A critical review", *Encephale*, Vol. 34, No. 3, March 2008, pp. 233 – 239.

tive Affectivity Schedule，*PANAS*）。[1] 该量表由 20 个形容词组成，其中代表积极情绪的形容词 10 个，代表消极情绪的形容词 10 个，被试要在 5 点量表（1 代表"非常轻微或根本没有"，5 代表"非常强烈"）上评定自己当前的情绪体验。后来，邱林等人对 PANAS 进行了修订，修订后的 PANAS 量表中代表积极情绪的形容词包括：兴奋的、自豪的、欣喜的、活跃的、感激的、快乐的、充满热情的、兴高采烈的、精力充沛的；代表消极情绪的形容词包括：难过的、紧张的、羞愧的、恐惧的、害怕的、内疚的、恼怒的、战战兢兢的、易怒的。[2]（2）情感强度量表（The Affect Intesity Measure，AIM）。该量表用于测验个体所体验的情绪强度，包括积极情绪、消极情绪的强度，平静度以及消极反应因素。[3]（3）情绪愉悦度和唤醒度评价。Bradley 和 Lang 在国际情绪图片系统中使用此评价方法作为标准，在愉悦度和唤醒度两个水平上进行评价，1 代表的程度最低，9 代表的程度最高，比如，在愉悦度上，1 为最不愉快，9 为最愉快；在唤醒度上，1 为非常平静，9 为非常强烈。[4]

二　情绪与反刻板印象、内隐性别刻板印象的关系

目前，尽管有许多研究探讨反刻板印象与刻板印象的关系以及

①　Watson D.，Clark L. A.，& Tellegen A.，"Development and validation of brief measures of positive and negative affect：The PANAS scales"，*Journal of Personality and Social Psychology*，Vol. 54，No. 6，March 1988，pp. 1063 – 1070.

②　邱林、郑雪、王雁飞：《积极情感消极情感量表（PANAS）的修订》，《应用心理学》2008 年第 3 期。

③　Larsen R. J.，& Diener E.，"Affect intensity as an individual difference characteristic：A review"，*Journal of Research in Personality*，No. 21，May 1987，pp. 1 – 39.

④　Bradley M. M.，& Lang P. J.，"The International Affective Picture System（IAPS）in the study of emotion and attention"，in J. A. Coan & J. J. B. Allen（Eds.），*Handbook of Emotion Elicitation and Assessment*，New York：Oxford University Press，2007，pp. 29 – 46.

情绪与刻板印象的关系，但是鲜有研究将情绪作为背景变量来考察反刻板印象对刻板印象的影响。

已有研究发现，当个体的反刻板印象被激活时，其通达性高于刻板印象的通达性。[1][2][3] 根据 MACK 模型，积极情绪意味着环境是安全可靠的，此时依赖更具通达性的一般知识结构具有高度适应性，[4] 而消极情绪则意味着个体当前的环境存在问题，在这样的环境中，再依赖一般知识结构可能会使个体感到适应不良、不安全。因此，消极情绪的个体倾向于注重任务中的具体信息以帮助个体解决环境中的问题，而不依赖其一般知识结构。[5] 从这个角度看，当个体的反刻板印象被激活时，由于此时反刻板印象的通达性高于刻板印象，积极情绪下个体会依赖反刻板印象进行判断，因而表现出积极情绪下反刻板印象对刻板印象的阻碍作用；而在消极情绪下，个体更倾向于依赖任务中的具体信息进行信息加工，而不是仅仅依赖当前具有更高通达性的信息。因此，消极情绪下反刻板印象可能不会对内隐刻板印象造成影响。

---

① Dasgupta N. , & Asgari S. , "Seeing is believing: Exposure to counterstereotypic women leaders and its effect on the malleability of automatic gender stereotyping", *Journal of Experimental Social Psychology*, No. 40, November 2004, pp. 642 –658.

② Blair I. V. , & Banaji M. , "Automatic and controlled processes in stereotype priming", *Journal of Personality and Social Psychology*, No. 70, December 1996, pp. 1142 – 1163.

③ Blair I. V. , Ma J. E. , & Lenton A. P. , "Imaging stereotypes away: The moderation of implicit stereotypes through mental imagery", *Journal of Personality and Social Psychology*, No. 81, August 2001, pp. 828 – 841.

④ Bless H. , Clore G. L. , Schwarz N. , Golisano V. , Rabe C. , & Wölk M. , "Mood and use of scripts: Does a happy mood really lead to mindlessness?" *Journal of Personality and Social Psychology*, No. 71, March 1996, pp. 665 – 679.

⑤ Lambert A. J. , Khan S. R. , Lickel B. A. , & Fricke K. , "Mood and the correction of positive versus negative stereotypes", *Journal of Personality and Social Psychology*, No. 72, June 1997, pp. 1002 – 1016.

此外，Clore 和 Huntsinger 提出，情绪意味着头脑中任何具有通达性的信息的价值。①② 具体而言，积极情绪意味着当前具有通达性的信息是有意义的，消极情绪意味着当前具有通达性的信息没有意义。因此，与消极情绪相比，积极情绪下的个体更倾向于依赖当前具有通达性的信息进行判断。在启动个体的反刻板印象时，反刻板印象比刻板印象具有更高的通达性③④⑤；同时，关于刻板印象研究中的情绪加工原则（the affective processing principle）认为，对于当前具有通达性的信息，积极情绪促进其使用，而消极情绪则抑制其使用。

值得注意的是，尽管上述两种观点不尽相同，但仍可看出，信息的通达性在情绪影响认知的过程中起关键作用。Avramova 和 Stapel 的研究结论进一步证实了信息通达性的重要性。他们的研究表明，积极情绪下当前高度通达的信息会对个体产生同化效应（assimilation）；而消极情绪下当前高度通达的信息会对个体产生异化效应（contrast）。⑥ 可见，情绪通过信息的通达性影响个体的后续

---

① Clore G. L., & Huntsinger J. R., "How emotion informs judgement and regulates thought", *TRENDS in Cognitive Sciences*, Vol. 11, No. 9, May 2007, pp. 393 – 399.

② Clore G. L., & Huntsinger J. R., "How the object of affect guides its impact", *Emotion Review*, No. 1, April 2009, pp. 39 – 54.

③ Blair I. V., Ma J. E., & Lenton A. P., "Imaging stereotypes away: The moderation of implicit stereotypes through mental imagery", *Journal of Personality and Social Psychology*, No. 81, December 2001, pp. 828 – 841.

④ Dasgupta N., & Asgari S., "Seeing is believing: Exposure to counterstereotypic women leaders and its effect on the malleability of automatic gender stereotyping", *Journal of Experimental Social Psychology*, No. 40, August 2004, pp. 642 – 658.

⑤ Blair I. V., & Banaji M., "Automatic and controlled processes in stereotype priming", *Journal of Personality and Social Psychology*, No. 70, May 1996, pp. 1142 – 1163.

⑥ Avramova Y. R., & Stapel D. A., "Moods as spotlights: The influence of moods on accessibility effects", *Journal of Personality and Social Psychology*, No. 95, December 2008, pp. 542 – 554.

行为。由于启动反刻板印象后，个体的反刻板印象通达性高于刻板印象通达性，①②③ 因此，根据上述观点可以推测：在积极情绪下启动反刻板印象会使个体更为依赖反刻板信息做出反应；而在消极情绪下启动反刻板印象会使个体放弃反刻板信息。那么，与消极情绪下启动反刻板印象相比，积极情绪下启动反刻板印象更能抑制性别刻板印象，即情绪可能在反刻板与性别刻板印象之间起调节作用。

①　Dasgupta N. , & Asgari S. , "Seeing is believing: Exposure to counterstereotypic women leaders and its effect on the malleability of automatic gender stereotyping", *Journal of Experimental Social Psychology*, No. 40, December 2004, pp. 642 – 658.

②　Blair I. V. , & Banaji M. , "Automatic and controlled processes in stereotype priming", *Journal of Personality and Social Psychology*, No. 70, October 1996, pp. 1142 – 1163.

③　Blair I. V. , Ma J. E. , & Lenton A. P. , "Imaging stereotypes away: The moderation of implicit stereotypes through mental imagery", *Journal of Personality and Social Psychology*, No. 81, November 2001, pp. 828 – 841.

第 四 章

情绪、反刻板印象与内隐性别刻板印象关系的实验研究

目前关于反刻板印象抑制性别刻板印象，以及情绪影响刻板印象的理论与研究成果均为后续相关研究提供了良好的理论基础，同时也为后续研究指引了新的研究方向。但以往研究仍存在以下几点不足。

第一，以往考察反刻板印象对内隐性别刻板印象影响的研究仅涉及特质、职业领域，但很少有研究涉及学科领域。一些研究发现，即使在推理和言语能力上仅存在十分微小的性别差异[①]，人们也仍然认为，男性比女性在推理等方面表现得更为优秀，女性比男性在写作等方面表现得更为突出。[②③] 事实上，这种观念在儿童时期便已存在。比如，有研究表明，小学儿童认为男孩比女

[①] Hyde J. S. , Fennema E. , & Lamon S. J. , "Gender differences in mathematics performance: A meta-analysis", *Psychological Bulletin*, No. 107, May 1990, pp. 139 – 155.

[②] Kiefer A. K. , & Sekaquaptewa D. , "Implicit stereotypes, gender identification, and math performance: a prospective study of female math students", *Psychological Science*, Vol. 18, No. 1, June 2010, pp. 13 – 18.

[③] Nosek B. A. , & Banaji M. R. , "The go/no-go association task", *Social Cognition*, No. 19, December 2001, pp. 625 – 666.

孩更擅长算数和推理，女孩比男孩更擅长阅读和写作。[1][2] 父母和教师也有同样的观念。[3] 这种学科性别刻板印象的存在一方面导致两性在某些科目表现上的差异越来越大，比如，一些女性认为自己缺乏推理能力，因此逐渐对数学等科目失去信心，在相关测验中的表现越来越不如男性;[4] 另一方面也导致部分职业男女比例失调，比如，Smith 的研究发现，女性在对言语能力有较高要求的职业（如，文秘、图书管理员）中所占比例远高于男性，而在其他领域，尤其是需要数学或数学相关能力的领域（如，物理、机械工程）中所占比例远小于男性;[5] 并且，一些需要科学能力的机构更愿意接受男性而非女性，因而扼杀了许多女性接受科学教育、从事科学事业的热情与机会。[6] 由此可见，探讨反刻板印象对学科性别刻板印象的影响，对于促进个体全面发展以及改善社会中的不良择业观念具有重要的现实意义。

第二，尽管许多研究者探讨了反刻板印象与性别刻板印象的关系，通过抑制性别刻板印象激活的方式来减弱甚至消除性别刻板印

① Eccles J. S. , & Harold R. D. , "Parentschool involvement during the early adolescent years", *Teachers College Record*, No. 94, April 1993, pp. 568 – 587.

② Eccles J. S. , Jacobs J. E. , & Harold R. D. , "Gender role stereotypes, expectancy effects, and parents socialization of gender differences", *Journal of Social Issue*, No. 46, March 1990, pp. 183 – 201.

③ Lummis M. , & Steverson H. W. , "Gender differences in beliefs and achievement: A cross-cultural study", *Developmental Psychology*, No. 26, November 1990, pp. 254 – 263.

④ Spencer S. J. , Steele C. M. , & Quinn D. M. , "Stereotype threat and women's math performance", *Journal of Experimental Social Psychology*, No. 35, July 1999, pp. 4 – 28.

⑤ Smith E. R. , "The role of exemplars in social judgment", in L. L. Martin & A. Tesser (Eds.), *The Construction of Social Judgment*, Hillsdale, NJ: Erlbaum, 1992.

⑥ Benbow C. P. , & Arjimand O. , "Predictors of high academic achievement in mathematics and science by mathematically talented students: A longitudinal study", *Journal of Educational Psychology*, No. 82, May 1990, pp. 430 – 441.

象的不良影响，但这些研究均忽略了被试的情绪状态。也就是说，以往关于反刻板印象影响性别刻板印象的研究均持有一个潜在的假设，即，这些研究均认为被试的情绪是一致的。然而，个体的情绪瞬息万变，并且，以往大量研究也表明，不同的情绪对个体的认知加工有不同影响。因此，有必要考察在不同的情绪状态下，反刻板印象对刻板印象的影响是否不同。或者说，有必要考察情绪在二者之间是否存在调节作用。

　　第三，以往关于情绪影响认知的研究更多地选择自传式回忆法来诱发被试的情绪。[1][2] 有研究者认为，虽然这种方法能够有效诱发出想要的情绪，但仍存在一定的缺陷，比如，这种方法要求被试有意识的合作，这就可能会导致要求特征的出现。[3] 鉴于此，本研究将采用更为生动的视频材料诱发被试的情绪，以避免要求特征对研究结果产生影响。

第一节　中国大学生学科性别刻板印象的研究

一　研究目的与假设

（一）研究目的

考察中国大学生是否存在学科性别刻板印象，以及这种学科性

---

① Bodenhausen G. V., Kramer G. P., & Sasser K., "Happiness and stereotypic thinking in social judgment", *Journal of Personality & Social Psychology*, Vol. 66, No. 4, July 1994, pp. 621 – 632.

② Krauth-Gruber S., & Ric F., "Affect and stereotypic thinking: A test of the mood-and-general-knowledge model", *Personality and Social Psychology Bulletin*, No. 26, March 2000, pp. 1587 – 1597.

③ Westermann R., Spies K., & Stahl G., "Relative Effectiveness and Validity of Mood Induction Procedures: A Meta-Analysis", *European Journal of Social Psychology*, Vol. 26, No. 4, December 1996, pp. 557 – 580.

别刻板印象是否存在专业差异，从而为进一步深入研究做准备。

（二）研究假设

中国大学生在内隐联想测验中的 D 值显著大于 0，即中国大学生存在学科性别刻板印象，他们更倾向于将男性与理科专业相联系，将女性与文科专业相联系。

二　研究方法

（一）被试

随机选取 J 市某高校大学生 31 名，男女生基本各半，文科生与理科生基本各半，年龄范围为 18—20 岁（$M = 19.03$，$SD = 0.41$）。所有被试均裸视或矫正视力正常，母语为汉语，无阅读障碍。参加本实验将获得一个小礼物。

（二）研究工具和实验材料

1. 研究工具

本研究采用联想电脑，用于呈现刺激材料和记录被试反应结果。使用 E-Prime 编写实验程序。

2. 实验材料

本研究中内隐联想测验（IAT）的概念词和属性词如下：

（1）目标概念维度：根据相关研究，选取 16 个与性别有关的词语作为目标概念维度，其中，男性词与女性词各 8 个，如，与男性有关的词为"儿子""父亲"，与女性有关的词为"妈妈""姑娘"。[1]

---

① Nosek B. A., Banaji M. R., & Greenwald A. G., "Math = Male, Me = Female, Therefore Math ≠ Me", *Journal of Personality and Social Psychology*, Vol. 83, No. 1, March 2002, pp. 44 – 59.

（2）属性维度：从 2010 年 B 市普通高等学校招生专业目录中选取文科与理科专业各 20 个，让 38 名大学生（男女各半，不参加正式实验）从中挑出 13 个属于文科的专业和 13 个属于理科的专业。最后，从被试挑选出的 20 个专业中选出归类一致性为 100% 的文科专业 7 个，如，"文学""哲学"，理科专业 7 个，如，"数学""化学"，作为内隐联想测验的属性维度。

3. 实验设计

本研究中，自变量为被试的专业，因变量为被试在内隐联想测验中的反应时所转化的 D 值。

（三）实验方法与程序

1. 培训主试

本研究的主试均为发展与教育心理学专业硕士研究生。施测前对主试进行严格的培训，包括预先让主试熟悉材料及施测程序，统一指导语，讲解施测过程中的注意事项，并对施测过程中可能会出现的问题做详细的讲解，以保证施测的顺利进行。经过培训，主试均能达到施测要求，如：言语清晰、语速适中、不给被试任何暗示等。

2. 实验程序

实验以小组团体施测的方式进行。被试进入实验室后在电脑前坐好，每名主试负责向 3 名被试介绍实验并宣读指导语。告知被试这是一项考察反应速度与准确性的任务，因此要既快又准确地根据说明做出反应。

本研究采用 Greenwald 等人设计的内隐联想测验程序。[①] 内隐

---

① Greenwald A. G. ， Nosek B. A. ， & Banaji M. R. ， "Understanding and using the implicit association test：Ⅰ. An improved scoring algorithm"， *Journal of Personality and Social Psychology*， Vol. 85， No. 2， May 2003， pp. 197 - 216.

联想测验包括目标概念维度和属性维度,二者形成相容联合辨别任务(男性词/理科专业,女性词/文科专业)和相反联合辨别任务(女性词/理科专业,男性词/文科专业)。内隐联想测验程序共包括七个部分,每部分的任务分别为:第一部分要求被试对屏幕中间词语所属的性别进行分类,把属于"男性"的词语归为一类并按 E 键,把属于"女性"的词语归为一类并按 I 键;第二部分要求被试对屏幕中间词语的专业类别进行分类,把属于"理科"的词语归为一类并按 E 键,把属于"文科"的词语归为一类并按 I 键;第三部分为相容联合辨别任务,当屏幕中间的词语属于"男性"或者"理科"时,将其归于一类并按 E 键,当屏幕中间的词语属于"女性"或者"文科"时,将其归于一类并按 I 键;第四部分重复第三部分;第五部分是对第二部分的反转,同样是对屏幕中间的词语所属的专业进行分类,把属于"文科"的词语归为一类并按 E 键,把属于"理科"的词语归为一类并按 I 键;第六部分为相反联合辨别任务,当屏幕中间的词语属于"男性"或者"文科"时,将其归于一类并按 E 键,当屏幕中间的词语属于"女性"或者"理科"时,将其归于一类并按 I 键;第七部分重复第六部分(内隐联想测验具体程序内容见表 4 - 1)。

表 4 - 1　　　　　　　　内隐联想测验(IAT)程序

内容	测验顺序						
	1	2	3	4	5	6	7
任务描述	目标维度辨别	属性维度辨别	相容联合辨别	相容联合辨别	属性维度反向辨别	相反联合辨别	相反联合辨别

内容	测验顺序						
	1	2	3	4	5	6	7
操作举例	叔叔ᵃ	数学ᵃ	叔叔ᵃ	叔叔ᵃ	文学ᵃ	叔叔ᵃ	叔叔ᵃ
			数学ᵃ	数学ᵃ		文学ᵃ	文学ᵃ
	姐姐ᵇ	文学ᵇ	姐姐ᵇ	姐姐ᵇ	数学ᵇ	姐姐ᵇ	姐姐ᵇ
			文学ᵇ	文学ᵇ		数学ᵇ	数学ᵇ

注：a 按 E 键，b 按 I 键。

　　每一部分开始前屏幕上都有相应的说明，告知被试这一部分的任务内容以及按键操作方法。比如，在第一部分中，说明如下："屏幕中间会出现一个词语，当它属于男性时，请按 E 键，将其归于左边；当它属于女性时，请按 I 键，将其归于右边。若无疑问，请按空格键继续。"被试按空格键呈现 500ms 的空屏，然后屏幕中间呈现一个词语，被试根据该部分的要求按 E 键或 I 键进行反应。刺激消失或 2500ms 后进入下一个词语的分类测验。屏幕左上方和右上方始终显示本部分需要分类的类别标签，如在第一部分中，左上方的类别标签为"男性"，右上方的类别标签为"女性"。被试每做出一个反应都会有相应的正确率和平均反应时出现作为反馈。计算机自动记录被试的反应时。

（四）计分方法

　　根据 Grennwald，Nosek 和 Banaji 提出的关于内隐联想测验的计分方法[1]，对数据进行如下处理：（1）只分析第三、第四、第六、第七部分的数据；（2）删除超过 10000 ms 的数据；（3）如果一个被试小于 300 ms 的反应时占 10% 以上，则剔除这个被试；（4）分别计算第三和第六两个部分的总体标准差 $ST1$、第四和第七两个部分的总体标准差 $ST2$；（5）计算每个部分中正确反应的平均反应时；（6）对于错误的反应，用每个部分的平均反应时加上 600 ms 来代替其反应时；（7）在错误反应的数据被替换后，计算每个部分的新的平均反应时；（8）分别计算第六部分与第三部分的平均反应时之差 $M1$，第七与第四部分的平均反应时之差 $M2$；（9）用平均数之差除以标准差，即用 $M1$ 除以 $ST1$ 得到 $D1$，用 $M2$ 除以 $ST2$ 得到 $D2$；（10）将 $D1$ 与 $D2$ 平均，得到 D 值。将 D 值作为本研究主要的观测指标。

（五）数据整理与统计分析

采用 Excel 进行数据管理，采用 SPSS13.0 软件进行统计分析。

三　研究结果

首先，对被试在内隐联想测验中的错误率进行分析，以保证实验是建立在被试认真反应的基础之上的。发现所有被试的错误率均在 20% 以下。

其次，根据 Greenwald，Nosek 和 Banaji[2] 提出的关于内隐联

---

① Greenwald A. G., Nosek B. A., & Banaji M. R., "Understanding and using the implicit association test：Ⅰ. An improved scoring algorithm", *Journal of Personality and Social Psychology*, Vol. 85, No. 2, September 2003, pp. 197 - 216.

② Greenwald A. G., Nosek B. A., & Banaji M. R., "Understanding and using the implicit association test：Ⅰ. An improved scoring algorithm", *Journal of Personality and Social Psychology*, Vol. 85, No. 2, July 2003, pp. 197 - 216.

想测验的计分方法，对数据进行初步分析处理并转换为 D 值，因此所有被试的数据均有效，最终被试人数为 31 人。初步分析发现，被试的 D 值不存在显著的性别差异，因此本研究中的所有统计分析将不再考虑性别因素。

为考察中国大学生是否存在学科性别刻板印象，对被试在内隐联想测验中的 D 值是否显著大于 0 进行单侧 t 检验。结果发现（见表 4－2），从总体以及各专业来看，被试的 D 值均显著大于 0。这表明，被试在不相容联合辨别任务中的反应时显著地长于相容联合辨别任务中的反应时，即中国大学生存在学科性别刻板印象。

表 4－2　　　　　各专业被试在内隐联想测验中的 D 值

（$M \pm SD$）及单侧 t 检验结果

专业	n	$M \pm SD$	t 值
文科	16	0.916 ± 0.405	9.048***
理科	15	0.662 ± 0.663	3.868**
合计	31	0.793 ± 0.551	8.013***

注：* 表明与 0 存在差异的程度，** $P < 0.01$，*** $P < 0.001$。

为考察这种学科性别刻板印象是否存在专业差异，对文科生与理科生在内隐联想测验中的 D 值进行 t 检验。结果发现，文科生与理科生的 D 值不存在显著差异，$t = 1.299$，$P > 0.05$，因此，后续研究将不再考虑被试专业这个变量。总体来说，不论文科生还是理科生，均倾向于将男性与理科相联系，将女性与文科相联系。

四　讨论

本研究发现，不论从总体还是从各专业上看，中国大学生在内隐联想测验中的 D 值显著大于 0，也就是说，中国大学生存在学科性别刻板印象。并且，文理科大学生在内隐联想测验中的 D 值并不存在显著差异，即不论文科生还是理科生，均倾向于将男性与理科相联系，将女性与文科相联系。这可能是由于，人们普遍认为，在问题解决能力和发散思维上，男性优于女性。[①] 由于理科比文科涉及更多的问题解决与发散思维，并且被试认同了这种社会对两性的不同期望[②]，因此产生了学科性别刻板印象。

已有研究已证实了从个体发展早期便存在学科性别刻板印象。如，小学生认为男生比女生更擅长教学，女孩在写作方面表现更出色。这种观念不仅体现在小学生中，还体现在父母和教师身上。这就可能导致女性认为自己缺乏推理能力，且部分需要数理推理能力的机构排斥女性，造成部分女性的发展困境。总之，学科性别刻板印象不仅在学业成就上造成两性的差异，也使两性在日后的职业选择上有所不同。因此，有必要探讨削弱学科性别刻板印象的方法，以减弱甚至消除其不良影响，从而促进个体更好的发展。

第二节 反刻板印象对内隐性别刻板印象的影响

一 研究目的与假设

（一）研究目的

考察反刻板印象对内隐性别刻板印象产生的影响。

---

① Nosek B. A., Banaji M. R., & Greenwald A. G., "Math = Male, Me = Female, Therefore Math ≠ Me", *Journal of Personality and Social Psychology*, Vol. 83, No. 1, June 2002, pp. 44 – 59.

② 程红娟、方晓义、蔺秀云：《大学生社会支持的调查研究》，《中国临床心理学杂志》2005 年第 3 期。

（二）研究假设

对于启动反刻板印象的被试，其内隐联想测验的 D 值显著小于没有启动反刻板印象的被试的 D 值，即，反刻板印象能有效地减弱个体的内隐性别刻板印象。

二　研究方法

（一）被试

随机选取 J 市某高校大学生 59 名。所有被试均裸视或矫正视力正常，母语为汉语，无阅读障碍。

（二）研究工具和实验材料

1. 研究工具

本研究采用联想电脑，用于呈现刺激材料和记录被试反应结果。使用 E - Prime 编写实验程序。

2. 实验材料

（1）反刻板样例材料

从已有研究以及互联网上收集 15 名中外女性科学家（如，居里夫人）的照片和简介资料，这些女性科学家在数学、物理、化学等理科领域取得了卓越的成就。[1] 将这些女性科学家的照片与简介通过幻灯片的方式呈现给被试，共计 45 张幻灯片。

（2）内隐联想测验的实验材料

同研究一。

---

① Dasgupta N. , & Asgari S. , "Seeing is believing Exposure to counterstereotypic women leaders and its effect on the malleability of automatic gender stereotyping", *Journal of Experimental Social Psychology*, No. 40, June 2004, pp. 642 - 658.

3. 实验设计

在本研究中，自变量为实验条件（实验组：启动反刻板印象；控制组：不启动反刻板印象），因变量为被试在内隐联想测验中的反应时所转化的 D 值。

（三）实验方法与程序

1. 培训主试

本研究的主试均为发展与教育心理学专业硕士研究生。实验前对主试进行严格的培训，包括预先让主试熟悉材料及实验程序，统一指导语，讲解实验过程中的注意事项，并对实验过程中可能会出现的问题做详细的讲解，以保证实验的顺利进行。经过培训，主试均能达到实验要求，如，言语清晰、语速适中、不给被试任何暗示等。

2. 实验程序

该实验以小组团体施测的方式进行。在实验前将被试随机分为两组（实验组：启动反刻板印象；控制组：不启动反刻板印象）。被试进入实验室后在电脑前坐好，每名主试负责向3—5名被试介绍实验并宣读指导语，并告知被试他们接下来要相继完成两个任务。首先要进行一项阅读任务，这项任务是为另外一项研究筛选实验材料，因此需要仔细阅读。然后向被试展示幻灯片文件。向实验组的被试呈现介绍女性科学家的幻灯片文件，而控制组的被试所观看的幻灯片内容则是关于如何阅读一本书。两组被试观看幻灯片文件的时间相同。待被试阅读完毕，告知被试接下来要进行一项考察反应速度与准确性的任务，因此要既快又准确地根据说明进行反应。接着进行内隐联想测验，其实验流程同研究一。

3. 计分方法

同研究一。

（四）数据整理与统计分析

采用 Excel 进行数据管理，采用 SPSS13.0 软件进行统计分析。

三　研究结果

首先，对被试在内隐联想测验中的错误率进行分析，以保证实验是建立在被试认真反应的基础之上的。所有被试的错误率均在 20% 以下，因此所有数据均为有效数据。然后，根据 Greenwald，Nosek 和 Banaji 提出的关于内隐联想测验的计分方法[①]，对数据进行初步分析处理并转换为 D 值。

其次，将两种实验条件下被试在内隐联想测验中所得的 D 值进行 t 检验，以考察反刻板印象是否能有效抑制学科性别刻板印象。结果发现，启动反刻板组被试的 D 值（$M = 0.60$，$SD = 0.40$）显著小于不启动反刻板组被试的 D 值（$M = 1.15$，$SD = 0.30$），$t = 6.15$，$P < 0.001$。这表明，启动反刻板印象组被试的学科性别刻板印象显著弱于不启动反刻板印象组被试的学科性别刻板印象。

为了进一步考察启动反刻板印象是否能够消除被试的学科性别刻板印象，对启动反刻板印象组的 29 名被试在内隐联想测验中所得的 D 值是否显著大于 0 进行单侧 t 检验。结果发现，启动反刻板印象组被试的 D 值（$M = 0.60$，$SD = 0.40$）显著大于 0，$t = 8.10$，$P < 0.001$。这表明，先前启动的反刻板印象的确能在一定程度上减弱学科性别刻板印象，但并不能完全消除学科性别刻板印象。

---

① Greenwald A. G. , Nosek B. A. , & Banaji M. R. , "Understanding and using the implicit association test：Ⅰ. An improved scoring algorithm", *Journal of Personality and Social Psychology*, Vol. 85, No. 2, March 2003, pp. 197 – 216.

四 讨论

在本研究中，向一部分被试呈现在数学、物理、化学等理科专业中取得卓越成就的女性科学家的样例，以启动其反刻板印象，另一部分被试则没有阅读这些内容。然后，通过内隐联想测验考察所有被试的学科性别刻板印象。结果发现，启动反刻板印象组的被试在内隐联想测验中的 D 值显著小于控制组被试，并且，两组被试的 D 值均显著大于 0。这表明，启动反刻板印象能够有效减弱学科性别刻板印象，但却不能完全消除。这可能是由于：在语义网络系统中，刻板印象除了与刻板一致概念之间存在正向联结，还存在和与其不一致的特质概念之间的负向联结[1]，或者说，刻板印象与反刻板印象是同时存在的。当个体的反刻板印象被激活时，刻板印象与刻板不一致概念之间的联结（反刻板联结）要强于刻板印象与刻板一致概念之间的联结（刻板联结），因此表现出反刻板印象对内隐刻板印象的削弱作用。尽管启动反刻板印象能有效减弱内隐刻板印象，说明内隐刻板印象是可以控制的，但是却并不能被完全消除，这就意味着，内隐刻板印象在一定程度上具有顽固性。

第三节 反刻板印象对内隐性别刻板印象的影响：情绪的调节作用

一 研究目的与假设

（一）研究目的

考察不同情绪下反刻板印象对内隐性别刻板印象的影响。

---

① Dijksterhuis A., & van Knippenberg A., "The knife that cuts both ways: Facilitated and inhibited access to traits as a result of stereotype-activation", *Journal of Experimental Social Psychology*, No. 32, October 1996, pp. 271 – 288.

（二）研究假设

假设 1：与消极情绪下启动反刻板印象相比，积极情绪下启动反刻板印象更能抑制性别刻板激活，即情绪可能在反刻板与性别刻板印象之间起调节作用。

假设 2：对于没有启动反刻板印象的被试来说，积极情绪下内隐联想测验的 D 值要显著大于消极情绪下被试的 D 值，即，积极情绪状态下被试的内隐性别刻板印象强度要大于消极情绪状态下被试的内隐性别刻板印象强度，或者说，在没有启动反刻板的情况下，与消极情绪相比，积极情绪对内隐性别刻板印象起促进作用。

二　研究方法

（一）被试

随机选取 J 市某高校大学生 116 名。所有被试均裸视或矫正视力正常，母语为汉语，无阅读障碍。

（二）研究工具和实验材料

1. 研究工具

本研究采用联想电脑，用于呈现刺激材料和记录被试反应结果。使用 E-Prime 编写实验程序。

2. 实验材料

以往研究认为，中性情绪难以诱发，其诱发结果往往偏向于积极或消极，并非预期的中性情绪。并且，有研究者发现，不诱发情绪并不等同于中性情绪，参加实验的被试本身可能就带有某种情绪，因此不管诱发与否都不能保证产生中性情绪。[①] 因此，本研究仅选用积极、消极两种情绪，并没有设立中性情绪。

（1）情绪诱发材料

从已有研究中选取视频片段作为情绪诱发材料，这些材料经评定均能有效诱发相应的情绪体验。[②] 其中，《猫和老鼠》片段为积极情绪诱发材料，《东京审判》片段为消极情绪诱发材料。

（2）反刻板样例材料

同研究二。

（3）内隐联想测验材料

同研究一。

（4）情绪诱发效果评估材料

情绪诱发效果评估材料采用 Watson，Clark 和 Tellegen[③]编制、邱林等人[④]修订的积极情绪消极情绪量表（PANAS）。该量表包括积极情绪分量表（PA）和消极情绪分量表（NA）。修订后的积极情绪分量表由 9 个形容词组成，分别是兴奋的、自豪的、欣喜的、活跃的、感激的、快乐的、充满热情的、兴高采烈的、精力充沛的；消极情绪分量表由 9 个形容词组成，分别是难过的、紧张的、羞愧的、恐惧的、害怕的、内疚的、恼怒的、战战兢兢的、易怒的。要求被试在 5 点量表中回答，他们在多大程度上体验到了量表中所描述的情绪，1—5 依次表示"没有或者非常轻微""有一

---

①　Zenasni F.，& Lubart T. I.，"Effects of emotional state on creativity"，Current Psychology Letters：Behavior，*Brain and Cognition*，No. 2，March 2002，pp. 33 – 50.

②　初玉霞：《任务特点、认知风格对情绪与创造表现关系的影响》，博士学位论文，山东师范大学，2011 年。

③　Watson D.，Clark L. A.，& Tellegen A.，"Development and validation of brief measures of positive and negative affect：The PANAS scales"，*Journal of Personality and Social Psychology*，Vol. 54，No. 6，April 1988，pp. 1063 – 1070.

④　邱林、郑雪、王雁飞：《积极情感消极情感量表（PANAS）的修订》，《应用心理学》2008 年第 3 期。

点""中等程度""很强烈""非常强烈"，得分越高表明体验到的积极或消极情绪越强烈。该量表具有良好的内部一致性，其中，积极情绪分量表的内部一致性系数 Cronbach α 值为 0.85，消极情绪分量表的内部一致性系数 Cronbach α 值为 0.84。在本研究中，积极情绪分量表的内部一致性系数 Cronbach α 值为 0.96，消极情绪分量表的内部一致性系数 Cronbach α 值为 0.93。

3. 实验设计

本研究采用 2（是否启动反刻板：启动、不启动）×2（情绪条件：积极情绪、消极情绪）的被试间实验设计。因变量为被试在内隐联想测验中的反应时所转化的 D 值。

（三）实验方法与程序

1. 培训主试

本研究的主试均为发展与教育心理学专业硕士研究生。实验前对主试进行严格的培训，包括预先让主试熟悉材料及实验程序，统一指导语，讲解实验过程中的注意事项，并对实验过程中可能会出现的问题做详细的讲解，以保证实验的顺利进行。经过培训，主试均能达到实验要求，如：言语清晰、语速适中、不给被试任何暗示等。

2. 实验程序

该实验以小组团体施测的方式进行。在实验前将被试随机分为两组（积极情绪组/消极情绪组），将积极情绪组和消极情绪组的被试分别随机分为启动反刻板印象组和不启动反刻板印象组，这样，被试被随机分为 4 组，即积极情绪/启动反刻板印象组、消极情绪/启动反刻板印象组、积极情绪/不启动反刻板印象组、消极情绪/不启动反刻板印象组。

被试进入实验室后在电脑前坐好，每名主试负责向 3—5 名被试介绍实验并宣读指导语，告知被试他们要完成以下几个任务。首先，让被试观看视频短片，积极情绪组的被试观看的是《猫和老鼠》片段，消极情绪组的被试观看的是《东京审判》片段。其次，告诉被试下面要进行一项阅读任务，这项任务是为另外一项研究筛选实验材料，因此需要仔细阅读幻灯片文件。向启动反刻板印象组的被试呈现介绍女性科学家的幻灯片文件，而不启动反刻板印象组的被试所观看的幻灯片内容则是关于如何阅读一本书，两组被试观看幻灯片的时间一致。待被试阅读完毕，告知被试接下来要进行一项考察反应速度与准确性的任务，因此要既快又准确地根据说明进行反应。最后，进行内隐联想测验，内隐联想测验程序同研究一。内隐联想测验结束后，被试完成对情绪诱发的效果评估。

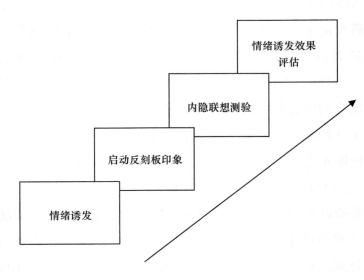

图 4 - 1　研究三实验流程

3. 计分方法

同研究一。

（四）数据整理与统计分析

采用 Excel 进行数据管理，采用 SPSS13.0 软件进行统计分析。

三　研究结果

（一）情绪诱发效果评估

将被试填写的积极情绪消极情绪量表得分进行统计，计算出被试在积极情绪分量表（PA）和消极情绪分量表（NA）上的总分。分别对两种情绪条件被试的 PA 和 NA 总分进行 t 检验，结果发现，在 PA 总分上，两种情绪条件的被试得分差异显著，积极情绪条件下的被试得分显著高于消极情绪条件下的被试得分，$t = 19.866$，$P < 0.001$；在 NA 总分上，两种情绪条件的被试得分差异也显著，消极情绪条件下的被试得分显著高于消极情绪条件下的被试得分，$t = -18.105$，$P < 0.001$。由此可知，本研究中情绪诱发是有效的。

（二）不同情绪条件下反刻板印象对内隐性别刻板印象的影响

首先对所有被试在内隐联想测验中的错误率进行分析，以确保实验建立在被试认真反应的基础之上。结果发现，有 3 名被试错误率高于 20%，因此删除这 3 名被试的所有数据。其次，根据 Greenwald，Nosek 和 Banaji[1] 提出的关于内隐联想测验的计分方法，对数据进行初步分析处理并转换为 D 值。最终，剩余 113 名

---

[1]　Greenwald A. G., Nosek B. A., & Banaji M. R., "Understanding and using the implicit association test：Ⅰ. An improved scoring algorithm", *Journal of Personality and Social Psychology*, Vol. 85, No. 2, June 2003, pp. 197–216.

被试的数据均有效。不同情绪条件下启动与不启动反刻板被试 D 值的平均数与标准差见表4 – 3。

表4 – 3　　　　不同情绪下启动与不启动反刻板印象
被试的 D 值平均数与标准差

		M	SD	n	有效数据人数
积极情绪	启动反刻板印象	0.56	0.37	26	
	不启动反刻板印象	0.91	0.28	29	
消极情绪	启动反刻板印象	0.88	0.23	27	
	不启动反刻板印象	0.71	0.29	31	
合计					113

为考察不同情绪条件下反刻板印象对内隐性别刻板印象的影响，对所有被试的 D 值进行2（是否启动反刻板）×2（情绪条件）的方差分析。结果发现，是否启动反刻板主效应不显著，F（1，109）= 2.395，P > 0.05；情绪条件主效应不显著，F（1，109）= 1.001，P > 0.05；交互作用（见图4 – 2）显著，F（1，109）= 21.120，P < 0.001。简单效应分析发现，一方面，积极情绪条件下，启动反刻板印象的被试的 D 值显著小于不启动反刻板印象被试的 D 值，t = – 3.911，P < 0.001；消极情绪下，启动反刻板印象的被试的 D 值显著大于不启动反刻板印象被试的 D 值，t = 2.429，P < 0.05。另一方面，积极情绪下启动反刻板的被试的 D 值显著小于消极情绪下启动反刻板的被试的 D 值，t = – 3.720，P < 0.001；对于不启动反刻板印象的被试，积极情绪下被试的 D 值显著大于消极情绪下被试的 D 值，t = – 3.689，P < 0.01。

这表明，一方面，在积极情绪状态下，先前启动的反刻板印象（与不启动反刻板印象相比）能够减弱性别刻板印象，但是，在消极情绪状态下，先前启动的反刻板印象（与不启动反刻板印象相比）反而增强性别刻板印象；不同情绪下启动反刻板印象的作用不同，与消极情绪下启动反刻板印象相比，积极情绪下启动反刻板印象更能抑制性别刻板印象。另一方面，在没有启动反刻板印象的条件下，积极情绪被试的性别刻板印象强度大于消极情绪被试的性别刻板印象强度。

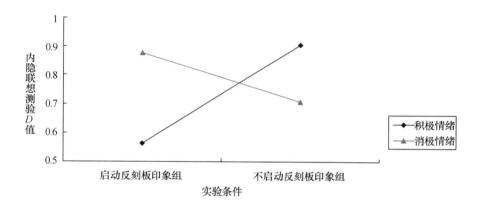

图4-2　不同情绪下启动与不启动反刻板印象被试的 D 值

四　讨论

本研究首先通过视频片段诱发被试的不同情绪体验，然后向其中一半被试呈现反刻板样例（在数学、物理等理科领域取得卓越成就的女性科学家），以启动其反刻板印象，最后所有被试完成考察学科性别刻板印象的内隐联想测验。结果发现，情绪条件与是否启动反刻板印象的交互效应显著。通过进一步分析表明，一方面，在没有启动反刻板印象的条件下，积极情绪被试的性别刻板印象强度大于消极情绪被试的性别刻板印象强度。这与以往研究相一致。比如，有研究发现，与消极情绪相比，积极情绪下的个体更倾向于依赖刻板印象做出判断。[1][2] 这可能是由于，积极情绪下，个体更关注整体信息，更倾向基于类别信息做出判断；而消极情绪下，个体更为注重具体信息，[3] 因此表现出积极情绪下的被试比消极情绪下的被试更依赖刻板印象做出反应。

另一方面，积极情绪下启动反刻板印象（与不启动反刻板印象相比）对刻板印象起削弱作用，这与假设相一致。这可能是由于：根据 MAGK 理论，积极情绪意味着当前环境是安全可靠的，个体倾向于利用其一般知识结构进行信息加工。而此时，由于反刻板印象的激活，反刻板印象比刻板印象更具通达性，因此，个体会依赖这种高度通达的反刻板印象进行后续的信息加工，从而表现为积极情绪下启动反刻板（与不启动反刻板相比）对刻板印象的削弱作用。而在消极情绪下，启动反刻板印象（与不启动反刻板印象相比）对刻板印象起促进作用。这可能是由于：消极情绪使个体放弃具有高度通达性的信息而专注于手头的任务。[4][5] 由于启动反刻板使其通达性高于性别刻板印象通达性，因此，消极

---

① Isbell L. M. , "Not all happy people are lazy or stupid: Evidence of systematic processing in happynmoods", *Journal of Experimental Social Psychology*, No. 40, October 2004, pp. 341 – 349.

② Bodenhausen G. V. , Kramer G. P. , & Sasser K. , "Happiness and stereotypic thinking in social judgment", *Journal of Personality & Social Psychology*, Vol. 66, No. 4, March 1994, pp. 621 – 632.

③ Huntsinger J. R. , Clore G. L. , & Bar-Anan Y. , "Mood and global-local focus: Priming a local focus reverses the link between mood and global-local processing", *Emotion*, Vol. 10, No. 5, March 2010, pp. 722 – 726.

④ Schwarz N. , & Clore G. L. , "Feelings and phenomenal experiences", in A. Kruglanski & E. T. Higgins (Eds.), *Social Psychology: Handbook of Basic Principles* (2nd ed.), New York: Guilford, 2007, pp. 385 – 407.

⑤ Bless H. , & Fiedler K. , "Mood and the regulation of information processing and behavior", in J. P. Forgas (Ed.), *Hearts and Minds: Affective Influences on Social Cognition and Behavior*, New York: Psychology Press, 2006, pp. 65 – 84.

情绪使其放弃反刻板信息，甚至使个体按照与反刻板信息相反的方向做出反应，[1][2] 即消极情绪启动反刻板印象反而促使个体按照刻板印象做出反应，从而产生异化效应（contrast）。

第四节　情绪与内隐性别刻板印象、反刻板印象的关系总讨论

一　中国大学生的学科性别刻板印象

研究一通过内隐联想测验考察中国大学生是否存在学科性别刻板印象。结果发现，所有被试均表现出明显的"IAT效应"，即他们在不相容联合辨别任务中的反应时显著地长于相容联合辨别任务中的反应时，这表明，中国大学生存在学科性别刻板印象，并且，这种学科性别刻板印象不存在性别与专业差异。也就是说，对于中国大学生，不论他们属于文科专业还是理科专业，不论他们是男生还是女生，都倾向于将男性与理科相联系，而将女性与文科相联系。以往有研究考察了大学生对于数学和艺术的内隐态度，得出了与本研究较为一致的结果。比如，Nosel，Banaji和Greenwald的研究发现，大学生普遍认为，男生更适合数学，而女生更适合艺术，这种态度同样不受被试性别与专业的影响。[3]这种态度的差异最终可能导致职业的不同：如，美国教育部门的

---

① Fishbach A. , & Labroo A. A. , "Be better or be merry: How mood affects self-control", *Journal of Personality and Social Psychology*, Vol. 93, No. 2, September 2007, pp. 158 – 173.

② Avramova Y. R. , & Stapel D. A. , "Moods as spotlights: The influence of moods on accessibility effects", *Journal of Personality and Social Psychology*, No. 95, July 2008, pp. 542 – 554.

③ Nosek B. A. , Banaji M. R. , & Greenwald A. G. , "Math = Male, Me = Female, Therefore Math ≠ Me", *Journal of Personality and Social Psychology*, Vol. 83, No. 1, December 2002, pp. 44 – 59.

一项统计发现，在计算机、数学、建筑工程等行业，男性远远多于女性。[1] 在中国也存在类似的现象，比如，高中文科班女生多，而理科班男生多，这种现象会一直延续到大学。

人们普遍认为，在问题解决能力和发散思维上，男性优于女性，[2] 而数学等理工类专业比文史类专业涉及更多的问题解决与发散思维。当人们认同这种社会对两性的不同期望[3]，便产生了学科性别刻板印象。有研究发现，这种学科性别刻板印象会影响女性在数学等领域的兴趣以及表现。[4][5][6][7][8] 因此，有必要深入探究削弱甚至消除这种不良刻板印象影响的方法，以促进个体更好

---

[1] U. S. Department of Education, National Center for Educational Statistics, National Assessment of Educational Progress, 1990, 1992, 1996, 2002, and 2003 Mathematics Assessments, Retrieved March 10, 2005, from < http: //nces. ed. gov/ > : 2005.

[2] Eccles J. S., & Harold R. D., "Parentschool involvement during the early adolescent years", *Teachers College Record*, No. 94, July 1993, pp. 568 – 587.

[3] 程虹娟、方晓义:《大学生社会支持的调查研究》,《中国临床心理学杂志》2005 年第 3 期。

[4] Davies P. G., Spencer S. J., Quinn D. M., & Gerhardstein R., "Consuming images: How television commercials that elicit stereotype threat can restrain women academically and professionally", *Personality and Social Psychology Bulletin*, No. 28, October 2002, pp. 1615 – 1628.

[5] Jacobs J. E., & Eccles J. S., "The impact of mothers' gender-role stereotypic beliefs on mothers' and children's ability perceptions", *Journal of Personality and Social Psychology*, Vol. 63, No. 6, September 1992, pp. 932 – 944.

[6] Quinn D. M., & Spencer S. J., "The interference of stereotype threat with women's generation of mathematical problem-solving strategies", *Journal of Social Issues*, No. 57, March 2001, pp. 55 – 71.

[7] Sekaquaptewa D., Espinoza P., Thompson M., vargas P., & von Hippel W., "Stereotypic explanatory bias: Implicit stereotyping as a predictor of discrimination", *Journal of Experimental Social Psychology*, No. 39, May 2003, pp. 75 – 82.

[8] Spencer S. J., Steele C. M., & Quinn D. M., "Stereotype threat and women's math performance", *Journal of Experimental Social Psychology*, No. 35, March 1999, pp. 4 – 28.

的发展。

二　反刻板印象对内隐性别刻板印象的影响

在研究二中，首先向一半被试呈现一些女性科学家的信息，包括照片以及文字介绍，她们在数学、物理等领域取得了卓越的成就，而另一半被试则会阅读一段介绍怎样读书的材料。接着所有被试完成考察内隐性别刻板印象的内隐联想测验。结果发现，呈现反刻板样例（女性科学家）的被试，其内隐联想测验的 D 值显著小于另一半被试，也就是说，反刻板印象能减弱内隐性别刻板印象。这与以往研究相一致。

这可能是由于：一方面，信息的通达性影响个体的判断。Smith 等人认为，内隐态度（如，内隐刻板印象）是背景依赖性结构（context-dependent constructions），知觉者对客体（如，某一社会群体）的评价和判断依赖于记忆中样例（如，社会群体的成员）的通达性：样例通达性越高，个体对它越依赖。[1][2] 由于启动反刻板使其通达性高于刻板印象通达性[3][4][5]，因此，个体更倾向于依赖反刻板信息（"女性擅长理科"）做出反应，从而表现为反

---

① Smith E. R., "The role of exemplars in social judgment", in L. L. Martin & A. Tesser (Eds.), *The Construction of Social Judgment*, Hillsdale, NJ: Erlbaum, 1992.

② Smith E. R., & Zárate M. A., "Exemplar-based model of social judgment", *Psychological Review*, Vol. 99, No. 1, July 1992, pp. 3 – 21.

③ Blair I. V., Ma J. E., & Lenton A. P., "Imaging stereotypes away: The moderation of implicit stereotypes through mental imagery", *Journal of Personality and Social Psychology*, No. 81, June 2001, pp. 828 – 841.

④ Dasgupta N., & Asgari S., "Seeing is believing Exposure to counterstereotypic women leaders and its effect on the malleability of automatic gender stereotyping", *Journal of Experimental Social Psychology*, No. 40, October 2004, pp. 642 – 658.

⑤ Blair I. V., & Banaji M., "Automatic and controlled processes in stereotype priming", *Journal of Personality and Social Psychology*, No. 70, December 1996, pp. 1142 – 1163.

刻板印象对学科性别刻板印象的抑制作用。另一方面,对反刻板角色的观察影响个体的判断。Eagly 等人的社会角色理论(social role theory)认为,性别刻板印象的习得和保持是通过观察男性和女性占据不同的社会角色形成的。[①] 当个体意识到男性和女性占据反刻板角色时,性别刻板印象可能会发生改变。[②③] 在本研究中,被试通过阅读材料对"女性科学家"这种反刻板角色有所了解,并意识到并非只有男性能在理科领域取得卓越成就,女性也可以在理科领域表现得十分出色,这样,被试的学科性别刻板印象减弱,从而表现出反刻板印象对学科性别刻板印象的抑制作用。

　　本研究还发现,尽管反刻板印象能够减弱学科性别刻板印象,但学科性别刻板印象仍然存在,并没有被完全消除。那么,性别刻板印象是否是无法消除的呢? Dasgupta 和 Asgari 的研究给我们很好的启示。他们选取女校学生(女生)和普通大学的女生作为被试考察其职业性别刻板印象。结果发现,两所学校的被试在入学时均存在职业性别刻板印象,即认为男性比女性更适合领导者的角色。一年后,采用与一年前同样的实验材料和程序再次考察两所学校被试的职业性别刻板印象。[④] 结果表明,普通大学的被

　　① Eagly A. H. , & Steffen V. J. , "Gender stereotypes stem from the distribution of women and men into social roles", *Journal of Personality and Social Psychology*, No. 46, November 1984, pp. 735 – 754.

　　② Eagly A. H. , & Steffen V. J. , "Gender stereotypes stem from the distribution of women and men into social roles", *Journal of Personality and Social Psychology*, No. 46, September 1984, pp. 735 – 754.

　　③ Diekman A. B. , & Eagly A. H. , "Stereotypes as dynamic constructs: Women and men of the past, present, and future", *Personality and Social Psychology Bulletin*, No. 26, August 2000, pp. 1171 – 1188.

　　④ Dasgupta N. , & Asgari S. , "Seeing is believing Exposure to counterstereotypic women leaders and its effect on the malleability of automatic gender stereotyping", *Journal of Experimental Social Psychology*, No. 40, June 2004, pp. 642 – 658.

试仍存在职业性别刻板印象，而女校的被试并没有显示出职业性别刻板印象。Dasgupta 和 Asgari 对此进行了深入分析，他们发现，被试接触女性领导者的频次对其职业性别刻板印象有重要影响。具体而言，被试接触女性领导者越频繁，其职业性别刻板印象越弱。由于女校的领导者全为女性，因此，女校被试接触女性领导者的机会比普通大学的被试更多、更频繁。通过这种长期的观察与接触，女校的被试逐渐意识到女性同样也能胜任领导者的角色。根据社会角色理论，当人们意识到男性和女性逐渐占据反刻板角色时，性别刻板印象会发生改变。① 因此，女校的被试在一年后并未显示出职业性别刻板印象。当然，Dasgupta 和 Asgari 研究结论的推广性有待探讨，并且，并非所有人都有条件长期接触反刻板角色，但我们仍可发现，社会环境对于性别刻板印象的减弱甚至消除起着重要作用。

三　反刻板印象对内隐性别刻板印象的影响：情绪的调节作用

　　尽管目前有许多研究考察内隐刻板印象及其与反刻板印象的关系，但这些研究均忽略了个体的情绪。个体的情绪瞬息万变，不同的情绪对个体的信息加工产生不同的影响。这些影响表现在问题解决②、刻板印象③等领域。尽管这些研究涉及的领域不同，

---

① Eagly A. H. , & Steffen V. J. , "Gender stereotypes stem from the distribution of women and men into social roles", *Journal of Personality and Social Psychology*, No. 46, May 1984, pp. 735 – 754.

② Schwarz N. , & Skurnik I. , "Feeling and thinking: implications or problem solving", in Davidson J. and Sternberg R. , eds. , *The Nature of Problem Solving*, Cambridge University Press, 2003, pp. 263 – 292.

③ Bless H. , Clore G. L. , Schwarz N. , Golisano V. , Rabe C. , & Wölk M. , "Mood and use of scripts: Does a happy mood really lead to mindlessness?" *Journal of Personality and Social Psychology*, No. 71, March 1996, pp. 665 – 679.

但仍得到了较为一致的结论，即，处于积极情绪状态（如，高兴）的个体更倾向于采用整体的、类别水平的加工方式，而处于消极情绪状态（如，悲伤）的个体更倾向于局部的、项目水平的加工方式。研究三在研究二的基础上，以个体的情绪为背景，进一步考察在不同情绪下反刻板印象与性别刻板印象之间的关系。实验中，向一半被试播放诱发积极情绪的视频片段，向另一半被试播放诱发消极情绪的视频片段。接着将每一种情绪条件下的被试随机分为两组，一组被试阅读关于女性科学家的材料，以激活其反刻板印象，而另一组被试则阅读其他材料。随后，所有被试完成考察学科性别刻板印象的内隐联想测验。

如前所述，情绪通过信息的通达性影响个体后续的信息加工①②：积极情绪促进个体利用高度通达的信息，而消极情绪使个体放弃这种信息而专注于手头的任务。由于启动反刻板印象使其通达性高于性别刻板印象通达性③④⑤，因此，积极情绪促进个体利用反刻板信息完成后续任务，即产生同化效应（assimilation）；

① Schwarz N., & Clore G. L., "Feelings and phenomenal experiences", in A. Kruglanski & E. T. Higgins (Eds.), *Social Psychology*: *Handbook of Basic Principles* (2nd ed.), New York: Guilford, 2007, pp. 385 – 407.

② Bless H., & Fiedler K., "Mood and the regulation of information processing and behavior", in J. P. Forgas (Ed.), *Hearts and Minds*: *Affective Influences on Social Cognition and Behavior*, New York: Psychology Press, 2006, pp. 65 – 84.

③ Blair I. V., Ma J. E., & Lenton A. P., "Imaging stereotypes away: The moderation of implicit stereotypes through mental imagery", *Journal of Personality and Social Psychology*, No. 81, June 2001, pp. 828 – 841.

④ Dasgupta N., & Asgari S., "Seeing is believing Exposure to counterstereotypic women leaders and its effect on the malleability of automatic gender stereotyping", *Journal of Experimental Social Psychology*, No. 40, October 2004, pp. 642 – 658.

⑤ Blair I. V., & Banaji M., "Automatic and controlled processes in stereotype priming", *Journal of Personality and Social Psychology*, No. 70, December 1996, pp. 1142 – 1163.

而消极情绪使个体放弃反刻板信息，甚至使个体按照与反刻板信息相反的方向做出反应①②，即产生异化效应（contrast）。因此，积极情绪下启动反刻板印象（与不启动反刻板印象相比）使性别刻板印象减弱；消极情绪下启动反刻板印象（与不启动反刻板印象相比）反而使性别刻板印象增强。综合这两方面的结果可以得出，与消极情绪下启动反刻板印象相比，积极情绪下启动反刻板印象更能抑制性别刻板印象，即，情绪在反刻板印象与性别刻板印象之间起调节作用。

　　此外，本研究还发现，在不启动反刻板印象的条件下，积极情绪个体的性别刻板印象强于消极情绪个体的性别刻板印象。这与以往研究结果相一致。如，Krauth-Gruber 和 Ric 在其研究中通过模拟法庭审判的场景考察情绪对刻板印象的影响。③ 结果发现，与消极情绪的被试相比，积极情绪的被试更依赖刻板印象判定"犯罪嫌疑人"是否有罪。目前，许多研究者从不同角度揭示了情绪影响刻板印象的机制。有研究者认为，情绪向个体传达了当前环境是否安全的信号：积极情绪代表个体当前所处环境是安全可靠的④，因此，依赖一般知识结构（如图式，刻板印象）进行

---

① Avramova Y. R. , & Stapel D. A. , "Moods as spotlights：The influence of moods on accessibility effects", *Journal of Personality and Social Psychology*, No. 95, June 2008, pp. 542－554.

② Fishbach A. , & Labroo A. A. , "Be better or be merry：How mood affects self-control", *Journal of Personality and* Social Psychology, Vol. 93, No. 2, July 2007, pp. 158－173.

③ Krauth-Gruber S. , & Ric F. , "Affect and stereotypic thinking：A test of the mood-and-general-knowledge model", *Personality and Social Psychology Bulletin*, No. 26, May 2000, pp. 1587－1597.

④ Bless H. , Schwarz N. , & Wieland R. , "Mood and the impact of category membership and individuating information", *European Journal of Social Psychology*, No. 26, June 1996, pp. 935－959.

判断具有高度适应性，并且有助于个体节省认知资源；而消极情绪意味着当前环境存在问题，此时还依赖一般知识结构进行判断是适应不良的表现，应该注意并依赖更加具体的信息以解决环境中的问题。Gasper 和 Clore 认为，积极情绪使个体倾向于采用类别水平的加工方式，而消极情绪使个体倾向于使用项目水平的加工方式。[1] 这样，与消极情绪的个体相比，积极情绪的个体更关注目标刺激的整体特点，从而更依赖刻板印象进行信息加工。尽管这些研究者解释情绪影响刻板印象的角度不同，但从总体上看可以发现，情绪影响个体的加工深度。具体而言，积极情绪使个体进行表面的、简单的信息加工，而消极情绪使个体进行较深层次的、精细的信息加工。[2][3][4][5][6] 正如 Avramova 和 Stapel 所比喻的那样：当你悲伤时，你只能看到一颗树木；当你开心时，你将看到

---

① Gasper K. , & Clore G. L. , " Attending to the big picture: Mood and global vs. local processing of visual information", *Psychological Science*, No. 13, October 2002, pp. 34 – 40.

② Bless H. , & Fiedler K. , " Mood and the regulation of information processing and behavior", in J. P. Forgas (Ed.), *Hearts and Minds: Affective Influences on Social Cognition and Behavior*, New York: Psychology Press, 2006, pp. 65 – 84.

③ Bless H. , Schwarz N. , & Wieland R. , " Mood and the impact of category membership and individuating information", *European Journal of Social Psychology*, No. 26, September 1996, pp. 935 – 959.

④ Fiedler K. , " Affective influences on social information processing ", in J. P. Forgas (Ed.), *Handbook of Affect and Social Cognition*, Mahwah, NJ: Erlbaum, 2001, pp. 163 – 185.

⑤ Martin C. L. , " Mood as input: A configural view of mood effects ", in L. L. Martin and G. L. Clore (Eds.), *Theories of Mood and Cognition: A User's Guidebook*, Mahwah, NJ: Erlbaum, 2001, pp. 135 – 157.

⑥ Schwarz N. , & Clore G. L. , " Feelings and phenomenal experiences ", in A. Kruglanski & E. T. Higgins (Eds.), *Social Psychology: Handbook of Basic Principles* (2nd ed.), New York: Guilford, 2007, pp. 385 – 407.

整片森林。①

四　本研究的局限及未来研究建议

本研究在考察中国大学生学科性别刻板印象的基础上，进一步探讨了反刻板印象对学科性别刻板印象的影响，以及情绪在二者之间的调节作用。但是，由于时间、精力和研究经验的局限，仍然存在许多不足之处。

第一，本研究的情绪诱发只选取了两个视频片段，相应地诱发了积极与消极两种情绪。但是，本研究仅在情绪的效价水平上探讨情绪对个体认知的影响，并未将具体情绪（如，愉悦、愤怒、悲伤）分离出来，分别探讨每一种具体情绪对认知的影响。那么，每一种具体的情绪对学科性别刻板印象有怎样的影响？特别是，这些具体情绪又是如何调节反刻板印象与学科性别刻板印象之间的关系？未来研究有必要通过诱发更为具体的情绪，以深入探讨具体情绪对个体认知的影响。

第二，本研究在综合前人研究的基础上，考察了情绪在反刻板印象与学科性别刻板印象之间的调节作用。但是，由于不同性别图式的个体在完成性别相关任务时存在差异，因此，未来研究有必要考察这种情绪的调节作用是否具有个体差异，从而为丰富情绪影响认知的理论提供更为有利的依据。

第三，尽管已有研究表明中性情绪难以诱发，并且，不诱发也不等同于中性情绪②，但是，由于缺乏基线条件，本研究不能

----

① Avramova Y. R. , & Stapel D. A. , "Moods as spotlights: The influence of moods on accessibility effects", *Journal of Personality and Social Psychology*, No. 95, December 2008, pp. 542 – 554.

② Zenasni F. , & Lubart T. I. , "Effects of emotional state on creativity", *Current Psychology Letters: Behavior, Brain and Cognition*, No. 2, May 2002, pp. 33 – 50.

更为深入地分析启动反刻板印象在不同情绪下的具体作用。因此, 未来研究有必要采用更为恰当的方法, 为不同情绪影响个体的认知提供一定的基线条件, 从而更为深入地探讨不同情绪的具体作用。

参考文献

一　中文文献

蔡华俭：《Greenwald 提出的内隐联想测验介绍》，《心理科学进展》2003 年第 3 期。

曹仁艳：《儿童性别刻板印象的发展与性别恒常性的关系：母亲教养态度的调节》，硕士学位论文，山东师范大学，2010 年。

程虹娟、方晓义、蔺秀云：《大学生社会支持的调查研究》，《中国临床心理学杂志》2005 年第 3 期。

初玉霞：《任务特点、认知风格对情绪与创造表现关系的影响》，博士学位论文，山东师范大学，2011 年。

连淑芳：《刻板印象的自动过程研究新进展》，《心理科学》2003 年第 1 期。

梁宁建、吴明证、高旭成：《基于反应时范式的内隐社会认知研究方法》，《心理科学》2003 年第 2 期。

庞小佳、张大均、王鑫强、王金良：《刻板印象干预策略研究述评》，《心理科学进展》2011 年第 2 期。

秦启文、余华：《性别角色刻板印象的调查》，《心理科学》2001 年第 5 期。

邱林、郑雪、王雁飞：《积极情感消极情感量表（PANAS）的修订》，《应用心理学》2008 年第 3 期。

战欣：《儿童性别刻板印象的发展及其对社会判断的影响》，硕士学位论文,山东师范大学,2006 年。

郑璞、刘聪慧、俞国良:《情绪诱发方法述评》,《心理科学进展》
2012 年第 1 期。

二 英文文献

Aboud, F. E., *Children and Prejudice*, New York: Basil Blackwell, 1988.

Aboud, F. E., & Amato, M., "Developmental and socialization influ-
ences on intergroup bias", in R. Brown & S. Gaertner (Eds.),
Blackwell Handbook in Social Psychology: Vol. 4: Intergroup Processes,
New York: Blackwell, 2001.

Abrams, D., & Hogg, M. A. (Eds.), *Social Identity Theory: Construc-
tive and Critical Advances*, New York: Springer-verlag, 1990.

Allport, G. W., *The Nature of Prejudice*, Reading, MA: Addison-
Wesley, 1954.

Amodio, D. M., Bartholow, B. D., & Ito, T. A., "Tracking the dynamics of the social brain: ERP approaches for social cognitive and affective neuroscience", *Social Cognitive and Affective Neuroscience*,
No. 9, 2014.

Amodio, D. M., Harmon-Jones, E., & Devine, P. G., "Individual
differences in the activation and control of affective race bias as assessed by startle eyeblink responses and self-report", Journal of Personality and Social Psychology, No. 84, 2003.

Amodio, D. M., Harmon-Jones, E., Devine, P. G., Curtin, J. J.,
Hartley, S. L., & Covert, A. E., "Neural signals for the detection of
unintentional race bias", *Psychological Science*, No. 15, 2004.

Anderson, N., & Armstead, C., "Toward understanding the association of socioeconomic status and health: A new challenge for the bio-

psychosocial approach", *Psychosomatic Medicine*, No. 57, 1995.

Archer, J., "Sex differences in social behavior: Are the social role and evolutionary explanations compatible?" *American Psychologist*, No. 51, 1996.

Aronson, J., Lustina, M. J., Good, C., Keough, K., Steele, C. M., & Brown, J., "When white men can't do math: Necessary and sufficient factors in stereotype threat", *Journal of Experimental Social Psychology*, No. 35, 1999.

Asuncion, A. G., & Lam, W. F., "Affect and impression formation: Influence of mood on person memory", *Journal of Experimental Social Psychology*, No. 31, 1995.

Avramova, Y. R., & Stapel, D. A., "Moods as spotlights: The influence of moods on accessibility effects", *Journal of Personality and Social Psychology*, No. 95, 2008.

Bach, P. B., Cramer, L. D., Warren, J. L., & Begg, C. B., "Racial differences in the treatment of early-stage lung cancer", *New England Journal of Medicine*, Vol. 341, No. 16, 1999.

Banaji, M. R., "Implicit stereotyping in person judgment", *Journal of Exerimental Social Psychology*, Vol. 65, No. 2, 1993.

Banaji, M. R., & Greenwald, A. G., "Implicit stereotype and unconscious prejudice", in Zanna, M. P., Olson, J. M. ed., *The Psychology of prejudice: The Ontario Symposium*, No. 7, 1994.

Banaji, M. R., & Hardin, C. D., "Automatic stereotyping", *Psychological Science*, No. 7, 1996.

Banaji, M. R., Nosek, B. A., & Greenwald, A. G., "No place for nostalgia in science: A response to Arkes and Tetlock", *Psychological Inquiry*, No. 15, 2004.

Bandura, A. , *Social Foundations of Thought and Action: A Social Cogntive Theory*, Englewood Cliffs, NJ: Prentice-Hall, 1986.

Bandura, A. , *Self-efficacy: The Exercise of Control*, New York: W. H. Freeman, 1997.

Bandura, A. , Ross, D. , & Ross, S. A. , "A comparative test of the status envy, social power, and secondary reinforcement theories of identificatory learning", *Journal of Abnormal and Social Psychology*, No. 67, 1963.

Banks, K. H. , Kohn-Wood, L. P. , & Spencer, M. , "An examination of the African American experience of everyday discrimination and symptoms of psychological distress", *Community Mental Health Journal*, No. 42, 2006.

Banos, R. , Liano, V. , & Botella, C. , "Changing Induced Moods Via Virtual Reality", Jsselsteijn, W. I. , *Lecture Notes in Computer Science: Persuasive Technology*, Heidelberg: Spriger Berlin, 2006.

Bargh, J. A. , Chen, M. , & Burrows, L. , "Automaticity of social behavior: Direct effects of trait construct and stereotype activation on action", *Journal of Personality & Social Psychology*, Vol. 71, No. 2, 1996.

Bargh, J. , "The cognitive monster: The case against the controllability of automatic stereotype effects", in S. Chaiken & Y. Trope (Eds.), *Dual-process Theories in Social Psychology*, New York: Guilford, 1999.

Bartholow, B. D. , Pearson, M. A. , Dickter, C. L. , Fabiani, M. , Gratton, G. , & Sher, K. H. , "Strategic control and medial frontal negativity: Beyond errors and response conflict", *Psychophysiology*, No. 42, 2005.

Baron, A. S. , & Banaji, M. R, "The development of implicit attitudes: Evidence of race evaluations from ages 6, 10 & adulthood", *Psychological Science*, No. 17, 2006.

Baumeister, R. F. , Faber, J. E. , & Wallace, H. M. , "Coping and ego-depletion: Recovery after the coping process", in C. R. Snyder (Ed.), *Coping: The Psychology of What Works*, New York: Oxford University Press, 1999.

Baumgartner, T. , Esslen, M. , & Jancke, L. , "From emotion perception to experience: Emotions evoked by pictures and classic music", *Inernational Journal of Psychophysiology*, Vol. 60, No. 1, 2006.

Beall, A. E. , & Sternberg, R. J. (Eds.), *The Psychology of Gender*, New York: Guilford Press, 1993.

Bem, S. L. , "Gender schema theory: A cognitive account of sex typing", *Psychological Review*, No. 88, 1981.

Benbow, C. P. , & Arjimand, O. , "Predictors of high academic achievement in mathematics and science by mathematically talented students: A longitudinal study", *Journal of Educational Psychology*, No. 82, 1990.

Berger, J. , Rosenholtz, S. J. , & Zelditch, M. , "Status organizing processes", *Annual Review of Sociology*, No. 6, 1980.

Berscheid, E. , "Forward", in A. E. Beall & R. J. Sternberg (Eds.), *The Psychology of Gender*, New York: Guilford Press, 1993.

Best, D. L. , Williams, J. E. , Cloud, J. M. , Davis, S. W. , Robertson, L. S. , Edwards, J. R. , Giles, H. , & Fowles, J. , "Development of Sex-Trait Stereotypes among Young Children in the United States, England, and Ireland", *Child Development*, Vol. 48, No. 4, 1977.

Bhui, K. , Stansfeld, S. , McKenzie, K. , Karlsen, S. , Nazroo, J. , & Weich, S. , "Racial/ Ethnic discrimination and common mental disorders among workers: Findings from the EMPIRIC study of ethnic minority groups in the United Kingdom", *American Journal of Public Health*, No. 95, 2005.

Biernat, M. , & Fuegen, K. , "Shifting standards and the evaluation of competence: Complexity in gender-based judgment and decision making", *Journal of Social Issues*, Vol. 57, No. 4, 2001.

Biernat, M. , & Vescio, T. K. , "She swings, she hits, she's great, she's benched: Implications of gender-based shifting standards for judgment and behavior", *Personality and Social Psychology Bulletin*, Vol. 28, No. 1, 2002.

Bigler, R. , "The role of classification skill in moderating environmental influences on children's gender stereotyping: A study of the functional use of gender in the classroom", *Child Development*, No. 66, 1995.

Bigler, R. , & Liben, L. , "Cognitive mechanisms in children's gender stereotyping: Theoretical and educational implications of a cognitive-based intervention", *Child Development*, No. 63, 1992.

Blair, I. V. , & Banaji, M. , "Automatic and controlled processes in stereotype priming", *Journal of Personality and Social Psychology*, No. 70, 1996.

Blair, I. V. , Ma, J. E. , & Lenton, A. P. , "Imaging stereotypes away: The moderation of implicit stereotypes through mental imagery", *Journal of Personality and Social Psychology*, No. 81, 2001.

Blakemore, J. E. O. , & Larue, A. A. , & Olejnik, A. B. , "Sex-appropriate toy preferences and the ability to conceptualize toys as sex-role related", *Developmental Psychology*, No. 15, 1979.

Bless, H. , Bohner, G. , Schwarz, N. , & Strack, F. , "Mood and persuasion: A cognitive response analysis", *Personality and Social Psychology Bulletin*, No. 16, 1990.

Bless, H. , Clore, G. L. , Schwarz, N. , Golisano, V. , Rabe, C. , & Wölk, M. , "Mood and use of scripts: Does a happy mood really lead to mindlessness?" *Journal of Personality and Social Psychology*, No. 71, 1996.

Bless, H. , & Fiedler, K. , "Mood and the regulation of information processing and behavior", in J. P. Forgas (Ed.), *Hearts and Minds: Affective Influences on Social Cognition and Behavior*, New York: Psychology Press, 2006.

Bless, H. , Schwarz, N. , & Kemmelmeier, M. , "Mood and stereotyping: Affective states and the use of general knowledge structures", in W. Stroebe & M. Hewstone (Eds.), *European Review of Social Psychology*, Chichester, UK: Wiley, Vol. 7, 1996.

Bless, H. , Schwarz, N. , & Wieland, R. , "Mood and the impact of category membership and individuating information", *European Journal of Social Psychology*, No. 26, 1996.

Bodenhausen, G. V. , "Stereotypes as judgmental heuristics: Evidence of circadian variations in discrimination", *Psychological Science*, No. 1, 1990.

Bodenhausen, G. V. , "Emotions, arousal, and stereotypic judgments: A heuristic model of affect and stereotyping", in D. M. Mackie & D. L. Hamilton (Eds.), *Affect, Cognition, and Stereotyping: Interactive Processes in Group Perception*, San Diego, CA: Academic Press, 1993.

Bodenhausen, G. V. , Kramer, G. P. , & Sasser, K. , "Happiness and stereotypic thinking in social judgment", *Journal of Personality & Social Psychology*, Vol. 66, No. 4, 1994.

Bodenhausen, G. V., & Macrae, C. N., "Stereotype activation and inhibition", in R. S. Wyer Jr. (Ed.), *Advances in Social Cognition*, Mahwah, NJ: Erlbaum, Vol. 11, 1998.

Bodenhausen, G. V., Schwarz, N., Bless, H., & Wanke, M., "Effects of atypical exemplars on racial beliefs: Enlightened racism or generalized appraisals?" *Journal of Experimental Social Psychology*, No. 31, 1995.

Bodenhausen, G. V., Sheppard, L. A., & Kramer, G. P., "Negative affect and social judgment: The differential impact of anger and sadness", *European Journal of Social Psychology*, No. 24, 1994.

Bodenhausen, G. V., & Wyer, R. S., "Effects of stereotypes on decision making and information-processing strategies", *Journal of Personality and Social Psychology*, No. 48, 1985.

Bosson, J. K., Haymovitz, E. L., & Pinel, E. C., "When saying and doing diverge", *Journal of Experimental Social Psychology*, No. 40, 2004.

Bradley, M. M., & Lang, P. J., "The International Affective Picture System (IAPS) in the study of emotion and attention", in J. A. Coan & J. J. B. Allen (Eds.), *Handbook of Emotion Elicitation and Assessment*, New York: Oxford University Press, 2007.

Branscombe, N. R., Schmitt, M. T., & Harvey, R. D., "Perceiving pervasive discrimination among African Americans: Implications for group identification and well-being", *Journal of Personality & Social Psychology*, Vol. 77, No. 1, 1999.

Brewer, M. B., "A dual process model of impression formation", in

T. K. Srull & R. S. Wyer（Eds.），*Advances in Social Cognition*，Hill-sdale，NJ：Erlbaum，Vol. 1，1988.

Brewer，D.，& Doughtie，E. B.，"Induction of Mood and Mood Shift"，*Journal of Clinical Psychology*，Vol. 36，No. 1，1980.

Brewer，M. B.，Dull，L.，& Lui，L.，"Perceptions of the elderly：Stereotypes as prototypes"，*Journal of Personality and Social Psychology*，No. 41，1981.

Briggs，S. R.，Cheek，J. M.，& Buss，A. H.，"Other directedness questionnaire"，*Journal of Personality and Social Psychology*，No. 38，1980.

Brody，G. H.，Chen，Y.，Murry，V. M.，Ge，X.，Simons，R. L.，Gibbons，F. X.，et al.，"Perceived discrimination and the adjustment of African American youths：A five-year longitudinal analysis with contextual moderation effects"，*Child Development*，No. 77，2006.

Brook，J. S.，Brook，D. W.，Balka，E. B.，& Rosenberg，G.，"Predictors of rebellious behavior in childhood：Parental drug use，peers，school environment，and child personality"，*Journal of Addictive Diseases*，No. 25，2006.

Bush，G.，Luu，P.，& Posner，M. I.，"Cognitive and emotional influences in anterior cingulate cortex"，*Trends in Cognitive Sciences*，No. 4，2000.

Buss，D. M.，"Psychological sex differences：Origins through sexual selection"，*American Psychologist*，No. 50，1985.

Buss，D. M.，& Schmitt，D. P.，"Sexual strategies theory：An evolutionary perspective on human mating"，*Psychological Review*，No. 100，1993.

Carter，D. B.，& Levy，G. D.，"Cognitive aspects of children's early sex-role development: The influence of gender schemas on preschoolers memories and preferences for sex-typed toys and activities", *Child Development*, No. 59, 1988.

Casagrande, S. S., Gary, T. L., LaVeist, T. A., Gaskin, D. J., & Cooper, L. A., "Perceived discrimination and adherence to medical care in a racially integrated community", *Journal of General Internal Medicine*, Vol. 22, No. 3, 2007.

Chaiken, S., Liberman, A., & Eagly, A. H., "Heuristic and systematic information processing within and beyond the persuasion context", in J. S. Uleman & J. A. Bargh (Eds.), *Unintended Thought*, New York: Guilford, 1989.

Chasteen, A. L., Kang, S. K., & Remedios, J. D., "Aging and stereotype threat: Development, process, and interventions", in M. Inzlicht & T. Schmader (Eds.), *Stereotype Threat: Theory, Process, and Application*, New York: Oxford University Press, 2011.

Chen, M., & Bargh, J. A., "Consequences of automatic evaluation: Immediate behavioral predispositions to approach or avoid the stimulus", *Personality and Social Psychology Bulletin*, No. 25, 1999.

Clore, G. L., & Huntsinger, J. R., "How emotion informs judgement and regulates thought", *TRENDS in Cognitive Sciences*, Vol. 11, No. 9, 2007.

Clore, G. L., & Huntsinger, J. R., "How the object of affect guides its impact", *Emotion Review*, No. 1, 2009.

Collins, E. C., Crandall, C. S., & Biernat, M., "Stereotypes and implicit social comparison: Shifts in comparison-group focus", *Journal of Experimental Social Psychology*, Vol. 42, No. 4, 2006.

Corning, A. F., "Self-esteem as a moderator between perceived discrimination and psychological distress among women", *Journal of Counseling Psychology*, Vol. 49, No. 1, 2002.

Correll, J., Park, B., Judd, C. M., & Wittenbrink, B., "The police officer's dilemma: Using ethnicity to disambiguate potentially threatening individuals", *Journal of Personality and Social Psychology*, No. 83, 2010.

Cottrell, C. A., Richards, D. A. R., & Nichols, A. L., "Predicting policy attitudes from general prejudice versus specific intergroup emotions", *Journal of Experimental Social Psychology*, No. 46, 2002.

Cota, A. A., & Dion, K. L., "Salience of gender and sex composition of ad hoc groups: An experimental test of distinctiveness theory", *Journal of Personality and Social Psychology*, Vol. 50, No. 4, 1986.

Cox, W. T., Abramson, L. Y., Devine, P. G., & Hollon, S. D., "Stereotypes, prejudice, and depression: The integrated perspective", *Perspectives on Psychological Science*, Vol. 7, No. 5, 2012.

Crandall, C. S., & Eshleman, A., "A justification-suppression model of the expression and experience of prejudice", *Psychological Bulletin*, Vol. 129, No. 3, 2003.

Crisp, R. J., Bache, L. M., & Maitner, A. T., "Dynamics of social comparison in counterstereotypic domains: Stereotype boost, not stereotype threat, for women engineering majors", *Social Influence*, No. 4, 2009.

Crocker, J., Major, B., & Steele, C. M., "Social stigma", in D. Gilbert, S. T. Fiske, & G. Lindzey (Eds.), *The Handbook of Social Psychology*, (4th ed.), Boston: McGraw Hill, Vol. 2, 1998.

Crocker, J., Voelkl, K., Testa, M., & Major, B., "Social stigma: The affective consequences of attributional ambiguity", *Journal of Personality and Social Psychology*, No. 60, 1991.

Crocker, J., & Wolfe, C. T., "Contingencies of self-worth", *Psychological Review*, 2001.

Croizet, J. C., & Millet, M., "Social class and test performance: From stereotype threat to symbolic violence and vice versa", in M. Inzlicht & T. Schmader (Eds.), *Stereotype Threat: Theory, Process, and Application*, New York: Oxford University Press, 2011.

Crosby, F., Bromley, S., & Saxe, L., "Recent unobtrusive studies of black and white discrimination and prejudice: A literature review", *Psychological Bulletin*, No. 87, 1980.

Cuddy, A. J., Fiske, S. T., & Glick, P., "Warmth and competence as universal dimensions of social perception: The stereotype content model and the BIAS map", *Advances in Experimental Social Psychology*, No. 40, 2008.

Cunningham, W., Preacher, K., & Banaji, M., "Implicit attitude measures: Consistency, stability, and convergent validity", *Psychological Science*, No. 12, 2001.

Czopp, A. M., "When is a compliment not a compliment? Evaluating expressions of positive stereotypes", *Journal of Experimental Social Psychology*, Vol. 44, No. 2, 2008.

Dasgupta, N., & Asgari, S., "Seeing is believing Exposure to counterstereotypic women leaders and its effect on the malleability of automatic gender stereotyping", *Journal of Experimental Social Psychology*, No. 40, 2004.

Dasgupta, N., & Greenwald, A. G., "On the malleability of automatic attitudes: Combating automatic prejudice with images of admired and disliked individuals", *Journal of Personality and Social Psychol-*

ogy, No. 81, 2001.

Dasgupta, N., McGhee, D. E., Greenwald, A. G., & Banaji, M. R., "Automatic preference for White Americans: Eliminating the familiarity explanation", *Journal of Experimental Social Psychology*, Vol. 36, No. 3, 2000.

Davis, S. K., Liu, Y., Quarells, R. C., Din-Dzietharn, R., & M. A. H. D. S. Group, "Stress-related racial discrimination and hypertension likelihood in a population-based sample of African Americans: The Metro Atlanta Heart Disease Study", *Ethnicity and Disease*, Vol. 15, No. 4, 2005.

Davies, P. G., Spencer, S. J., Quinn, D. M., & Gerhardstein, R., "Consuming images: How television commercials that elicit stereotype threat can restrain women academically and professionally", *Personality and Social Psychology Bulletin*, No. 28, 2002.

Davies, P. G., Spencer, S. J., & Steele, C. M., "Clearing the air: Identity safety moderates the effects of stereotype threat on women's leadership aspirations", *Journal of Personality and Social Psychology*, No. 88, 2005.

Deaux, K., & Major, B., "Putting gender into context: An interactive model of gender related behavior", *Psychological Review*, No. 94, 1987.

Deaux, K., Reid, A., Mizrahi, K., & Ethier, K. A., "Parameters of social identity", *Journal of Personality & Social Psychology*, Vol. 68, No. 2, 1995.

De Houwer, J., "Comparing measures of attitudes at the functional and procedural level: Analysis and implications", in R. E. Petty, R. H. Fazio, & P. Briñol (Eds.), *Attitudes: Insights from the new implicit measures*, Mahwah, NJ: Erlbaum, 2009.

De Houwer, J., "The extrinsic affective Simon task", *Experimental Psychology*, No. 50, 2003.

De Houwer, J., Beckers, T., & Moors, A., "Novel attitudes can be faked on the Implicit Association Test", *Journal of Experimental Social Psychology*, No. 43, 2007.

De Houwer, J., Custers, R., & De Clercq, A., "Do smokers have a negative implicit attitude toward smoking?" *Cognition and Emotion*, No. 20, 2006.

Devine, P. G., "Stereotypes and prejudice: Their automatic and controlled components", *Journal of Personality and Social Psychology*, No. 56, 1989.

Dewall, C. N., Baumeister, R. F., Stillman, T. F., & Gailliot, M. T., "Violence restrained: Effects of self-regulation and its depletion on aggression", *Journal of Experimental Social Psychology*, No. 43, 2007.

Diekman, A. B., & Eagly, A. H., "Stereotypes as dynamic constructs: Women and men of the past, present, and future", *Personality and Social Psychology Bulletin*, No. 26, 2000.

Dijksterhuis, A., & van Knippenberg, A., "The knife that cuts both ways: Facilitated and inhibited access to traits as a result of stereotype-activation", *Journal of Experimental Social Psychology*, No. 32, 1996.

Dovidio, J., Evans, N., & Tyler. R., "Racail stereotypes: The contents of their cognitive representation", *Journal of Experimental Social Psychology*, No. 22, 1986.

Dovidio, J. F., Gaertner, S. L., Isen, A. M., & Lowrance, R., "Group representations and intergroup bias: Positive affect, similarity, and group size", *Personality and Social Psychology Bulletin*, Vol. 21, No. 8, 1995.

Dovidio, J. F. , Kawakami, K. , Johnson, C. , Johnson, B. , & Howard, A. , "On the nature of prejudice: Automatic and controlled processes", *Journal of Experimental Social Psychology*, No. 33, 1997.

Dunham, Y. , & Degner, J. , "Origins of intergroup bias: Developmental and social cognitive research on intergroup attitudes", *European Journal of Social Psychology*, Vol. 40, No. 4, 2010.

Eagly, A. H. , *Sex Differences in Social Behavior: A Social Role Interpretation*, Hillsdale, NJ: Erlbaum, 1987.

Eagly, A. H. , "Reporting sex differences", *American Psychologist*, No. 42, 1987.

Eagly, A. H. , & Steffen, V. J. , "Gender stereotypes stem from the distribution of women and men into social roles", *Journal of Personality and Social Psychology*, No. 46, 1984.

Eccles, J. S. , & Harold, R. D. , "Parentschool involvement during the early adolescent years", *Teachers College Record*, No. 94, 1993.

Eccles, J. S. , Jacobs, J. E. , & Harold, R. D. , "Gender role stereotypes, expectancy effects, and parents' socialization of gender differences", *Journal of Social Issue*, No. 46, 1990.

Ellemers, J. , Spears, R. , & Doosje, B. (Eds.), *Social Identity: Context, Commitment, Content*, Oxford, UK: Blackwell, 1999.

Ellis, H. C. , & Ashbrook, P. W. , "Resource allocation model of the effects of depressed mood states on memory", in K. Fiedler & J. Forgas (Eds.), *Affect, Cognition, and Social Behaviour*, Toronto: Hogrefe, 1988.

Epstein, C. F. , *Deceptive distinctions: Sex, gender, and the social order*, New Haven, CT: Yale University Press, 1988.

Epstein, C. F. , "The multiple realities of sameness and difference: Ideology and practice", *Journal of Social Issues*, No. 53, 1997.

Facione, N. C. , & Facione, P. A. , "Perceived prejudice in healthcare and women's health protective behavior", *Nursing Research*, No. 56, 2007.

Fagot, B. I. , "Changes in thinking about early sex role development", *Developmental Review*, No. 5, 1985.

Fagot, B. I. , & Leinbach, M. D. , "The young child's gender schema: Environmental input, internal organization", *Child Development*, No. 60, 1989.

Fiedler, K. , "Affective influences on social information processing", in J. P. Forgas (Ed.), *Handbook of affect and social cognition*, Mahwah, NJ: Erlbaum, 2001, pp. 163 – 185.

Fishbach, A. , & Labroo, A. A. , "Be better or be merry: How mood affects self-control", *Journal of Personality and Social Psychology*, Vol. 93, No. 2, 2007.

Fiske, S. T. , "Examining the role of intent: Toward understanding its role in stereotyping and prejudice", in J. S. Uleman & J. A. Bargh (Eds.), *Unintended Thought*, New York: Guilford, 1989.

Fiske, S. T. , Cuddy, A. J. C. , Glick, P. , & Xu, J. A. , "model of (often mixed) stereotype content: Competence and warmth respectively follow from perceived status and competition", *Journal of Personality & Social Psychology*, Vol. 82, No. 6, 2002.

Fiske, S. T. , & Neuberg, S. L. , "A continuum of impression formation from category-based to individuating processes: Influences of information and motivation on attention and interpretation", in M. P. Zanna (Ed.), *Advances in experimental social psychology*, San Diego, CA: Academic Press, Vol. 23, 1990.

Forgas, J. P. , & Fiedler, K. , "Us and them: Mood effects on intergroup discrimination", *Journal of Personality and Social Psychology*, No. 70, 1996.

Freud, S. , *Three Contributions to the Theory of Sex*, New York: Nervous and Mental Disease Publishing Co. (original work published 1905) .

Freud, S. , "Introductory lectures on psychoanalysis", in J. Strachey (Ed.), *The Standard Edition of the Complete Psychological Works of Sigmund Frued*, London: Hogarth (original work published 1916): 1916/1963, Vol. 18.

Frijda, N. , "The laws of emotion", *American Psychologist*, No. 43, 1988.

Fritz, T. , Jentschke, S. , Gosselin, N. , Sammler, D. , Peretz, I. , Turner, R. , et al. , "Universal recognition of three basic emotions in music", *Current Biology*, No. 19, 2009.

Gaertner, S. L. , & Dovidio, J. F. , "Racism among the well intentioned", in E. Clausen & J. Bermingham (Eds.), *Pluralism, Racism and Public Policy: The Search for Equality*, Boston, MA: G. K. Hall, 1981.

Gaertner, S. L. , & Dovidio, J. F. , "The aversive form of racism", in S. L. Gaertner & J. F. Dovidio (Eds.), *Prejudice, Discrimination and Racism*, Orlando, FL: Academic: 1986.

Gardener, S. L. , & McLaughlin, J. P. , "Racial stereotypes: Ascialtions and ascriptions of positive and negative characteristics", *Social Psychology Quarterly*, No. 46, 1983.

Gasper, K. , & Clore, G. L. , "Attending to the big picture: Mood and global vs. local processing of visual information", *Psychological Science*, No. 13, 2002.

Gawronski, B. , & Bodenhausen, G. V. , "The associative-propositional evaluation model: Theory, evidence, and open questions", *Advances in Experimental Social Psychology*, No. 44, 2011.

Gawronski, B. , Deutsch, R. , Mbirkou, S. , Seibt, B. , & Strack, F. , "When 'just say no' is not enough: Affirmation versus negation training and the reduction of automatic stereotypes activation", *Journal of Experimental Social Psychology*, No. 44, 2008.

Gawronski, B. , & Payne, B. K. (Eds.), *Handbook of Implicit Social Cognition: Measurement, Theory, and Applications*, Guilford Press, 2010.

Gehring, W. J. , Goss, B. , Coles, M. G. , & Meyer, D. E. , "A neural system for error detection and compensation", *Psychological Science*, No. 4, 1993.

Geis, F. L. , "Self-fulfilling prophecies: A social psychological view of gender", in A. E. Beall & R. J. Sternberg (Eds.), *The Psychology of Gender*, New York: Guilford Press, 1993.

Gerson, K. , "Continuing controversies in the sociology of gender", *Sociological Forum*, No. 5, 1990.

Gibbons, F. X. , Yeh, H. C. , Gerrard, M. , Cleveland, M. J. , Cutrona, C. , Simons, R. L. , & Brody, G. H. , "Early experience with racial discrimination and conduct disorder as predictors of subsequent drug use: A critical period analysis", *Drug and Alcohol Dependence*, Vol. 88, No. 1, 2007.

Gilet, A. L. , "Mood induction procedures: A critical review", *En-*

cephale, Vol. 34, No. 3, 2008.

Gilbert, D. T., & Hixon, J. G., "The trouble of thinking: Activation and application of stereotypic beliefs", *Journal of Personality and Social Psychology*, No. 60, 1991.

Gilligan, C., *In a Different Voice*, Cambridge: Harvard University Press: 1982.

Glauser, A. S., "Legacies of racism", *Journal of Counseling & Development*, No. 77, 1999.

Gold, C., Voracek, M., & Wigram, T., "Effects of music therapy for children and adolescents with psychopathology: A meta-analysis", *Journal of Child Psychology and Psychiatry*, Vol. 45, No. 6, 2004.

Gonzales, P. M., Blanton, H., & Williams, K. J., "The effect of stereotype threat and double-minority status on the test performance of Latino women", *Personality and Social Psychology Bulletin*, No. 28, 2002.

Govorun, O., & Payne, B. K., "Ego-depletion and prejudice: Separating automatic and controlled components", *Social Cognition*, Vol. 24, No. 2, 2006.

Greenwald, A. G., "What cognitive representations underlie attitudes?", *Bulletin of the Psychonomic Society*, Vol. 28, No. 2, 1990.

Greenwald, A. G. & Banaji, M. R., "Implicit Social Cognition. Attitudes, Self-Esteem, and stereotypes", *Psychological Review*, Vol. 102, No. 1, 1995.

Greenwald, A. G., McGhee, D. E., & Schwartz, J. L. K., "Measuring individual differences in implicit cognition: The implicit association test", *Journal of Personality and Social Psychology*, Vol. 74, No. 6, 1998.

Greenwald, A. G., Nosek, B. A., & Banaji, M. R., "Understanding and using the implicit association test: Ⅰ. An improved scoring algorithm", *Journal of Personality and Social Psychology*, Vol. 85, No. 2, 2003.

Hamilton, D. L., & Sherman, J. W., "Stereotypes", in R. S. Wyer, Jr., & T. K. Srull (Eds.), *Handbook of Social Cognition* (2nd ed.), Hillsdale, NJ: Erlbaum, Vol. 2, 1994.

Hamilton, D. L., & Trolier, T. K., "Stereotypes and stereotyping: An overview of the cognitive approach", in J. F. Dovidio & S. L. Gaertner (Eds.), *Prejudice, Discrimination, and Racism*, New York: Academic Press, 1986.

Han, H. A., Olson, M. A., & Fazio, R. H., "The influence of experimentally-created extrapersonal associations on the Implicit Association Test", *Journal of Experimental Social Psychology*, No. 42, 2006.

Han, S., & Northoff, G., "Culture-sensitive neural substrates of human cognition: a transcultural neuroimaging approach", *Nature Reviews Neuroscience*, Vol. 9, No. 8, 2008.

Harrell, J. P., Hall, S., & Taliaferro, J., "Physiological responses to racism and discrimination: An assessment of the evidence", *American Journal of Public Health*, No. 93, 2003.

Harris, R., Tobias, M., Jeffreys, M., Waldegrave, K., Karlsen, S., & Nazroo, J., "Effects of self-reported racial discrimination and deprivation on Māori health and inequalities in New Zealand: Cross-sectional study", *Lancet*, Vol. 367, No. 9527, 2006.

Hetherington, E. M., "The effects of familial variables on sex typing, on parent-child similarity, and on imitation in children", in J. P. Hill (Ed.), *Minnesota Symposia on Child Psychology*, Minneapolis: University of Minnesota Press, Vol. 1, 1967.

Hewstone, M., "Revision and change of stereotypic beliefs: In search for the elusive subtyping model", in W. Stroebe & M. Hewstone (Eds.), *European Review of Social Psychology*, Chichester, UK: Wiley, Vol. 5, 1994.

Houben, K., & Wiers, R. W., "Are drinkers implicitly positive about drinking alcohol? Personalizing the alcohol-IAT to reduce negative extrapersonal contamination", *Alcohol and Alcoholism*, No. 42, 2007.

Hudley, C., & Graham, S., "Stereotypes of achievement striving among early adolescents", *Social Psychology of Education*, Vol. 5, No. 2, 2001.

Huntsinger, J. R., Clore, G. L., & Bar-Anan, Y., "Mood and global-local focus: Priming a local focus reverses the link between mood and global-local processing", *Emotion*, Vol. 10, No. 5, 2010.

Huston, A. C., "Sex typing", in P. H. Mussen (Series Ed.) & E-. M. Hetherington (Vol. Ed.), *Handbook of Child Psychology: Vol. 4. Socialization, Personality, and Social Development* (4th ed.), New York: Wiley, 1983.

Hyde, J. S., Fennema, E., & Lamon, S. J., "Gender differences in mathematics performance: A meta-analysis", *Psychological Bulletin*, No. 107, 1990.

Isbell, L. M., "Not all happy people are lazy or stupid: Evidence of systematic processing in happynmoods", *Journal of Experimental Social Psychology*, No. 40, 2004.

Inzlicht, M., Aronson, J., Good, C., & McKay, L., "A particular resiliency to threatening environments", *Journal of Experimental*

Social Psychology, No. 42, 2006.

Inzlicht, M. , & Gutsell, J. N. , "Running on empty: Neural signals for self-control failure", *Psychological Science*, No. 18, 2007.

Inzlicht, M. , & Kang, S. K. , "Stereotype threat spillover: How coping with threats to social identity affects aggression, eating, decision-making, and attention", *Journal of Personality and Social Psychology*, No. 99, 2010.

Inzlicht, M. , McKay, L. , & Aronson, J. , "Stigma as ego depletion: How being the target of prejudice affects self-control", *Psychological Science*, Vol. 17, No. 3, 2006.

Isen, A. M. , "Toward understanding the role of affect in cognition", in R. S. Wyer & T. K. Krull (Eds.), *Handbook of Social Cognition*, Hillsdale, NJ: Lawrence Erlbaum, Vol. 3, 1984.

Isen, A. M. , "Positive affect, cognitive processes, and social behaviour", in L. Berkowitz (Ed.), *Advances in Experimental Social Psychology*, New York: Academic Press, Vol. 20, 1987.

Jackson, J. W. , & Smith, E. R. , "Conceptualizing social identity: A new framework and evidence for the impact of different dimensions", *Personality & Social Psychology Bulletin*, Vol. 25, No. 1, 1999.

Jacobs, J. E. , & Eccles, J. S. , "The impact of mothers' gender-role stereotypic beliefs on mothers' and children's ability perceptions", *Journal of Personality and Social Psychology*, Vol. 63, No. 6, 1992.

Jacoby, L. L. , Kelley, C. M. , Brown, J. , & Jasechko, J. , "Becoming famous overnight. Limits on the ability to avoid unconscious influences of the past", *Journal of personality and Social Psychology*, Vol. 16, No. 2, 1989.

Johns, M. , Inzlicht, M. , & Schmader, T. , "Stereotype threat and executive resource depletion: Examining the influence of emotion regulation", *Journal of Experimental Psychology-General*, No. 137, 2008.

Jussim, L. , "Accuracy in social perception: Criticisms, controversies, criteria, components, and cognitive processes", *Advances in Experimental Social Psychology*, No. 37, 2005.

Jussim, L. , Cain, T. R. , Crawford, J. T. , Harber, K. , Cohen, F. , & Nelson, T. , "The unbearable accuracy of stereotypes", *Handbook of Prejudice*, *Stereotyping*, *and Discrimination*, 2009.

Kagan, J. , "The acquisition and significance of sex-typing and sex-role identity", in M. Hoffman & L. Hoffman (Eds.), *Review of Child Development Research*, New York: Russell Sage, Vol. 1, 1964.

Kahneman, D. , "A perspective on judgment and choice: Mapping bounded rationality", *American Psychologist*, No. 58, 2003.

Kaiser, C. R. , Major, B. , & McCoy, S. K. , "Expectations about the future and the emotional consequences of perceiving prejudice", *Personality & Social Psychology Bulletin*, Vol. 30, No. 2, 2004.

Karpinski, A. , & Hilton, J. L. , "Attitudes and the Implicit Association Test", *Journal of Personality and Social Psychology*, No. 81, 2001.

Katz, D. , & Braly, K. W. , "Racial stereotypes of one hundred college students", *Journal of Abnormal and Social Psychology*, No. 28, 1933.

Kawakami, K. , Dovidio, J. F. , Moll, J. , Hermsen, S. , & Russin, A. , "Just say no (to stereotyping): effects of training in the negation of stereotypic associations on stereotypic activation", *Journal of*

Personality and Social Psychology, No. 78, 2000.

Kay, A. C., Day, M. V., Zanna, M. P., & Nussbaum, A. D., "The insidious (and ironic) effects of positive stereotypes", *Journal of Experimental Social Psychology*, Vol. 49, No. 2, 2013.

Keller, J., "Blatant stereotype threat and women's math performance: Self-handicapping as a strategic means to cope with obtrusive negative performance expectations", *Sex Roles*, No. 47, 2022.

Keller, J., & Dauenheimer, D., "Stereotype threat in the classroom: Dejection mediates the disrupting threat effect on women's math performance", *Personality and Social Psychology Bulletin*, No. 29, 2003.

Kernis, M. H., Paradise, A. W., Whitaker, D. J., "Wheatman, S. R., & Goldman, B. N., Master of one's psychological domain? Not likely if one's self-esteem is unstable", *Personality and Social Psychology Bulletin*, No. 26, 2000.

Kessler, R. C., Mickelson, K. D., & Williams, D. R., "The prevalence, distribution, and mental health correlated of perceived discrimination in the United States", *Journal of Health and Social Behavior*, No. 40, 1999.

Khaylis, A., Waelde, L., & Bruce, E., "The role of ethnic identity in the relationship of race-related stress to PTSD symptoms among young adults", *Journal of Trauma & Dissociation*, No. 8, 2007.

Kiefer, A. K., & Sekaquaptewa, D., "Implicit stereotypes, gender identification, and math performance: a prospective study of female math students", *Psychological Science*, Vol. 18, No. 1, 2010.

Klonoff, E. A., Landrine, H., & Ullman, J. B., "Racial discrimination and psychiatric symptoms among blacks", *Cultural Diversity and Ethnic Minority Psychology*, Vol. 5, No. 4, 1999.

Koenig, A. M. , & Eagly, A. H. , "Stereotype threat in men on a test of social sensitivity", *Sex Roles*, No. 52, 2005.

Kohlberg, L. , "A cognitive-developmental analysis of children's sex-role concepts and attitudes", in E. E. Maccoby (Ed.), *The Development of Sex Differences*, Stanford, CA: Stanford University Press, 1966.

Konečni, V. J. , "Does music induce emotion? A theoretical and methodological analysis", *Psychology of Aesthetics, Creativity and the Arts*, No. 2, 2008.

Krauth-Gruber, S. , & Ric, F. , "Affect and stereotypic thinking: A test of the mood-and-general-knowledge model", *Personality and Social Psychology Bulletin*, No. 26, 2000.

Kray, L. J. , Reb, J. , Galinsky, A. D. , & Thomson, L. , "Stereotype reactance at the bargaining table: The effect of stereotype activation and power on claiming and creating value", *Personality and Social Psychology Bulletin*, No. 30, 2004.

Kubota, J. T. , Banaji, M. R. , & Phelps, E. A. , "The neuroscience of race", *Nature Neuroscience*, Vol. 15, No. 7, 2012.

Lambert, A. J. , Khan, S. R. , Lickel, B. A. , & Fricke, K. , "Mood and the correction of positive versus negative stereotypes", *Journal of Personality and Social Psychology*, No. 72, 1997.

Landrine, H. , & Klonoff, E. A. , "The schedule of racist events: A measure of racial discrimination and a study of its negative physical and mental health consequences", *Journal of Black Psychology*, No. 22, 1996.

Lassiter, G. D. , Koenig, L. J. , & Apple, K. J. , "Mood and behavior perception: Dysphoria can increase and decrease effortful process-ing of information", *Personality and Social Psychology Bulletin*, No. 22, 1996.

Larsen, R. J. , & Diener, E. , "Affect intensity as an individual differ-ence characteristic: A review", *Journal of Research in Personality*, No. 21, 1987.

Larson, A. , Gillies, M. , Howard, P. J. , & Coffin, J. , "It's e-nough to make you sick: The impact of racism on the health of Aborigi-nal Australians", *Australian and New Zealand Journal of Public Health*, Vol. 31, No. 4, 2007.

Leary, M. R. , Kelly, K. M. , Cottrell, C. A. , & Schreindorfer, L. S. , Individual differences in the need to belong: Mapping the no-mological network. Unpublished manuscript, Wake Forest University, 2006.

Lee, Y. T. , Jussim, L. J. , & McCauley, C. R. , *Stereotype Accura-cy: Toward Appreciating Group Differences*, Washington, DC: Ameri-can Psychological Association, 1995.

Levine, L. J. , & Burgess, S. L. , "Beyond general arousal: Effects of specific emotions on memory", *Social Cognition*, No. 15, 1997.

Levy, G. D. , & Fivush, R. , "Scripts and gender: A new approach for examining gender-role development", *Developmental Review*, No. 13, 1993.

Lewis, G. J. , Kandler, C. , & Riemann, R. , "Distinct heritable in-fluences underpin in-group love and out-group derogation", *Social Psychological and Personality Science*, Vol. 5, No. 4, 2014.

Lincoln, K. D. , Chatters, L. M. , Taylor, R. J. , & Jackson, J. S. , "Profiles of depressive symptoms among African Americans and Caribbean Blacks", *Social Science & Medicine*, No. 65, 2007.

Lippman, W., *Public Opinion*, New York: Harcourt & Brace, 1922.

Lobel, T. E., Gruber, R., Govrin, N., & Mashraki-Pedhatzur, S., "Children's gender-related inferences and judgments: A cross-cultural study", *Developmental Psychology*, Vol. 37, No. 6, 2001.

Logel, C., Iserman, E. C., Davies, P. G., Quinn, D. M., & Spencer, S. J., "The perils of double consciousness: The role of thought suppression in stereotype threat", *Journal of Experimental Social Psychology*, No. 45, 2009.

Logel, C., Peach, J., & Spencer, S. J., "Threatening gender and race: Different manifestations of stereotype threat", in M. Inzlicht & T. Schmader (Eds.), *Stereotype Threat: Theory, Process, and Application*, New York: Oxford University Press, 2011.

Lorber, J., *Paradoxes of Gender*, New Haven, CT: Yale University Press: 1994.

Lowery, B. S., Hardin, C. D., & Sinclair, S., "Social influence effects on automatic racial prejudice", *Journal of Personality and Social Psychology*, No. 81, 2001.

Luhtanen, R., & Crocker, J., "A collective self-esteem scale: Self-evaluation of one's social identity", *Personality and Social Psychology Bulletin*, No. 18, 1992.

Lummis, M., & Steverson, H. W., "Gender differences in beliefs and achievement: A cross-cultural study", *Developmental Psychology*, No. 26, 1990.

Luu, P., Collins, P., & Tucker, D. M., "Mood, personality and self-monitoring: Negative affect and emotionality in relation to frontal lobe mechanisms of error monitoring", *Journal of Experimental Psy-*

chology: *General*, No. 129, 2000.

Mackie, A., "The chemical basis of food detection in the lobster Homarus gammarus", *Marine Biology*, Vol. 21, No. 2, 1973.

Macrae, C. N., Bodenhausen, G. V., & Miline, A. B., "The dissection of selection in person perception: Inhibitory processes in social stereotyping", *Journal of Personality and Social Psychology*, Vol. 69, No. 3, 1995.

Mackie, D. M., & Hamilton, D. L. (Eds.), *Affect*, *Cognition*, *and Stereotyping*: *Interactive Processes in Group Perception*, San Diego, CA: Academic Press, 1993.

Mackie, D. M., & Worth, L. T., "Processing deficits and the mediation of positive affect in persuasion", *Journal of Personality and Social Psychology*, No. 57, 1989.

Macrae, C. N., Milne, A. B., & Bodenhausen, G. V., "Stereotypes as energy-saving devices: A peek inside the cognitive toolbox", *Journal of Personality and Social Psychology*, No. 66, 1994.

Madon, S., Jussim, L., Keiper, S., Eccles, J., Smith, A., & Palumbo, P., "The accuracy and power of sex, social class, and ethnic stereotypes: A naturalistic study in person perception", *Personality & Social Psychology Bulletin*, Vol. 24, No. 12, 1998.

Major, B., Kaiser, C. R., & McCoy, S. K., "It's not my fault: When and why attributions to prejudice protect self-esteem", *Personality & Social Psychology Bulletin*, Vol. 29, 2003.

Major, B., & Schmader, T., "Coping with stigma through psychological disengagement", in J. Swim, & C. Stangor (Eds.), *Prejudice*: *The Target's Perspective*, San Diego: Academic Press, 1998.

Maner, J. K., Kenrick, D. T., Becker, D. V., Robertson, T. E.,

Hofer, B., Neuberg, S. L., et al., "Functional projection: How fundamental social motives can bias interpersonal perception", *Journal of Personality & Social Psychology*, Vol. 88, No. 1, 2005.

Martin, C. L., "Children's use of gender-related information in making social judgments", *Developmental Psychology*, No. 25, 1991.

Martin, C. L., "New directions for investigating children's gender knowledge", *Developmental Review*, No. 13, 1993.

Martin, C. L., "Mood as input: A configural view of mood effects", in L. L. Martin and G. L. Clore (Eds.), *Theories of Mood and Cognition: A user's Guidebook*, Mahwah, NJ: Erlbaum, 2001.

Markus, H., Crane, M., Bernstein, S., & Siladi, M., "Self-schemas and gender", *Journal of Personality and Social Psychology*, No. 42, 1982.

Martin, C. L., "Stereotypes about children with traditional and nontraditional gender roles", *Sex Roles*, No. 33, 1995.

Martin, C. L., & Halverson, C. F., "A schematic processing model of sex typing and stereotyping in children", *Child Development*, No. 52, 1981.

Martin, C. L., & Little, J. K., "The relation of gender understanding to children's sex-typed preferences and gender stereotypes", *Child Development*, No. 61, 1990.

Martin, C. L., Ruble, D. N., & Szkrybalo, J., "Cognitive theories of early gender development", *Psychological Bulletin*, Vol. 128, No. 6, 2002.

Martin, C. L., Ward, D. W., Achee, J., & Wyer, R. S., "Mood as input: People have to interpret the motivational implications of their moods", *Journal of Personality and Social Psychology*,

No. 64, 1993.

Martin, J. K., Tuch, S. A., & Roman, P. M., "Problem drinking patterns among African Americans: The impacts of reports of discrimination, perceptions of prejudice, and 'risky' coping strategies", *Journal of Health and Social Behavior. Special Issue: Race, Ethnicity and Mental Health*, No. 44, 2003.

Masicampo, E. J., & Baumeister, R. F., "Toward a physiology of dual-process reasoning and judgment: Lemonade, willpower, and expensive rule-based analysis", *Psychological Science*, No. 19, 2008.

Mendes, W. B., & Jamieson., "Embodied stereotype threat: Exploring brain and body mechanisms underlying performance impairments", in M. Inzlicht & T. Schmader (Eds.), *Stereotype Threat: Theory, Process, and Application*, New York: Oxford University Press, 2011.

McCauley, C., & Stitt, C. L., "An individual and quantitative measure of stereotypes", *Journal of Personality and Social Psychology*, No. 36, 1978.

McConnell, A. R., & Leibold, J. M., "Relations between the Implicit Association Test, explicit racial attitudes, and discriminatory behavior", *Journal of Experimental Social Psychology*, No. 37, 2001.

McGlone, M. S., & Aronson, J., "Stereotype threat, identity salience, and spatial reasoning", *Journal of Applied Developmental Psychology*, No. 27, 2006.

Mulia, N., "Social disadvantage, stress, and alcohol use among Black, Hispanic, and White Americans: Findings from the 2005 U. S. national alcohol survey", *Journal of Studies on Alcohol and Drugs*, No. 69, 2008.

Murphy, M. C., Steele, C. M., & Gross, J. J., "Signaling threat: How situational cues affect women in math, science, and engineering settings", *Psychological Science*, No. 18, 2007.

Nussbaum, D. A., & Steele, C. M., "Situational disengagement and persistence in the face of adversity", *Journal of Experimental Social Psychology*, No. 43, 2007.

Neuberg, S. L., Kenrick, D. T., & Schaller, M., "Human threat management systems: Self-protection and disease avoidance", *Neuroscience & Biobehavioral Reviews*, Vol. 35, No. 4, 2011.

Newheiser, A., & Olson, K. R., "White and Black American children's implicit intergroup bias", *Journal of Experimental Social Psychology*, No. 48, 2012.

Nguyen, H. D., & Ryan, A. M., "Does stereotype threat affect test performance of minorities and women? A meta-analysis of experimental evidence", *Journal of Applied Psychology*, No. 93, 2008.

Nosek, B. A., & Banaji, M. R., "The go/no-go association task", *Social Cognition*, No. 19, 2001.

Nosek, B. A., Banaji, M. R., & Greenwald, A. G., "Math = Male, Me = Female, Therefore Math ≠ Me", *Journal of Personality and Social Psychology*, Vol. 83, No. 1, 2002.

Nosek, B. A., Greenwald, A. G., & Banaji, M. R., "Understanding and using the Implicit Association Test: II. Method variables and construct validity", *Personality and Social Psychology Bulletin*, No. 31, 2005.

Nosek, B. A., Smyth, F. L., Hansen, J. J., Devos, T., Lindner, N. M., Ranganath, K. A., et al., "Pervasiveness and correlates of implicit attitudes and stereotypes", *European Review of Social Psy-

chology, No. 16, 2008.

Oakes, P. J. , Haslam, S. A. , & Turner, J. C. , *Sterotyping and Social Reality*, Oxford, UK: Blackwell, 1994.

Oakes, P. J. , Turner, J. C. , & Haslam, S. A. , "Perceiving people as group members: The role of fit in the salience of social categorizations", *British Journal of Social Psychology*, No. 30, 1991.

Olson, M. A. , & Fazio, R. H. , "Reducing the influence of extra-personal associations on the Implicit Association Test: Personalizing the IAT", *Journal of Personality and Social Psychology*, No. 86, 2004.

Olson, M. A. , Fazio, R. H. , & Han, H. A. , "Conceptualizing personal and extrapersonal associations", *Social Psychology and Personality Compass*, No. 3, 2009.

Olson, J. M. , vernon, P. A. , Harris, J. A. , & Jang, K. L. , "The heritability of attitudes: A study of twins", *Journal of Personality and Social Psychology*, Vol. 80, No. 6, 2001.

Olsson, A. , & Phelps, E. A. (Eds.), *Understanding Social Evaluations: What We Can (and cannot) Learn from Neuroimaging*, in B. Wittenbrink & N. Schwartz (Eds.), *Implicit Measures of Attitudes*, New York: Guilford, 2007.

Osgood, C. E. , Suci, G. J. , & Tannenbaum, P. H. , *The Measurement of Meaning*, Urbana, IL: University of Illinois Press, 1957.

Oswald, F. L. , Mitchell, G. , Blanton, H. , Jaccard, J. , & Tetlock, P. E. , "Predicting ethnic and racial discrimination: A meta-analysis of IAT criterion studies", *Journal of Personality and Social Psychology*, No. 105, 2013.

Pascoe, E. A. , & Richman, L. , "Perceived discrimination and health: A metaanalytic review", *Psychological Bulletin*,

No. 135, 2009.

Pauker, K., Ambady, N., & Apfelbaum, E. P., "Race salience and essentialist thinking in racial stereotype development", *Child Development*, Vol. 81, No. 6, 2010.

Payne, B. K., "Prejudice and perception: the role of automatic and controlled processes in misperceiving a weapon", *Journal of Personality and Social Psychology*, No. 81, 2001.

Payne, B. K., Cheng, C. M., Govorun, O., & Stewart, B. D., "An inkblot for attitudes: Affect misattribution as implicit measurement", *Journal of Personality and Social Psychology*, Vol. 89, No. 3, 2005.

Payne, D. E., & Mussen, P. H., "Parent-child relations and father identification among adolescent boys", *Journal of Abnormal and Social Psychology*, No. 52, 1956.

Peretz, I., Gagnon, L., & Bouchard, B., "Music and emotion: perception determinants, immediacy, and isolation after brain damage", *Cognition*, Vol. 68, No. 2, 1998.

Perry, D. G., White, A. J., & Perry, L. C., "Does early sex typing result from children's attempts to match their behavior to sex role stereotypes?" *Child Development*, No. 55, 1984.

Quinn, D. M., & Spencer, S. J., "The interference of stereotype threat with women's generation of mathematical problem-solving strategies", *Journal of Social Issues*, No. 57, 2001.

Reyna, V. F., & Brainerd, C. J., "Fuzzy-trace theory: An interim synthesis", *Learning and Individual Differences*, Vol. 7, No. 1, 1995.

Riach, P. A., & Rich, J., "Fishing for discrimination", *Review of*

Social Economy, Vol. 62, No. 4, 2004.

Richards, J. M., & Gross, J. J., "Emotion regulation and memory: The cognitive costs of keeping one's cool", *Journal of Personality and Social Psychology*, No. 79, 2000.

Roberts, C. B., Vines, A. I., Kaufman, J. S., & James, S. A., "Cross-sectional association between perceived discrimination and hypertension in African American men and women: The Pitt County Study", *American Journal of Epidemiology*, Vol. 167, No. 5, 2007.

Roccas, S., & Brewer, M., "Social identity complexity", *Personality & Social Psychology Review*, Vol. 6, No. 2, 2002.

Rosenkrantz, P. S., Vogel, S. R., Bee, H. L., Broverman, I. K., & Broverman, D. M., "Sex-role stereotypes and self-concepts in college students", *Journal of Consulting and Clinical Psychology*, Vol. 32, No. 3, 1968.

Rowe, D. C., *The Limits of Family Influence: Genes, Experience, and Behavior*, New York: The Guilford Press, 1994.

Ruble, D., & Martin, C., "Gender development", in W. Damon (Ed.), *Handbook of Child Psychology* (5th ed.), New York: Wiley, 1998.

Ruble, D. N., Taylor, L. J., Cyphers, L., Greulich, F. K., Lurye, L. E., & Shrout, P. E., "The role of gender constancy in early gender development", *Child Development*, Vol. 78, No. 4, 2007.

Rudman, L. A., Ashmore, R. D., & Gary, M. L., "'Unlearning' automatic biases: The malleability of implicit stereotypes and prejudice", *Journal of Personality and Social Psychology*, No. 81, 2001.

Ryan, C., Park, B., & Judd, C., "Assessing stereotype accuracy: Implications for understanding the stereotyping process", in

C. N. Macrae, C. Stangor, & M. Hewstone (Eds.), *Stereotypes and Stereotyping*, *New York*: *Guilford*, 1996.

Rydell, R. J., & McConnell, A. R., "Understanding implicit and explicit attitude change: A systems of reasoning analysis", *Journal of Personality and Social Psychology*, No. 91, 2006.

Schacter, D. L., Chiu, C. Y. P., & Ochsner, K. N., "Implicit memory: A selective review", *Annual Review of Neuroscience*, No. 16, 1993.

Scheier, M. F., & Carver, C. S., "The Self-consciousness Scale: A revised version for use with general populations", *Journal of Applied Social Psychology*, No. 15, 1985.

Schmader, T., & Beilock, S., "An integration of processes that underlie stereotype threat", in M. Inzlicht & T. Schmader (Eds.), *Stereotype Threat*: *Theory*, *Process*, *and Application*, New York: Oxford University Press, 2011.

Schmader, T., Johns, M., & Forbes, C., "An integrated process model of stereotype threat effects on performance", *Psychological Review*, No. 115, 2008.

Schmitt, M. T., Branscombe, N. R., Kobrynowicz, D., & Owen, S., "Perceiving discrimination against one's gender group has different implications for well-being in women and men", *Personality and Social Psychology Bulletin*, Vol. 28, No. 2, 2002.

Schmitt, M. T., Spears, R., & Branscombe, N. R., "Constructing a minority group identity out of shared rejection: The case of international students", *European Journal of Social Psychology*, Vol. 33, No. 1, 2003.

Schneider, D. J., *The Psychology of Stereotyping*, New York: Guilford，2004.

Schwarz, N. , "Feeling as information: Informational and motivational functions of affective states", in E. T. Higgins & R. M. Sorrentino (Eds.), *Handbook of Motivation and Cognition*, New York: Guilford, Vol. 2, 1990.

Schwarz, N. , Bless, H. , & Bohner, G. , "Mood and persuasion: Affective states influence the processing of persuasive communications", in M. P. Zanna (Ed.), *Advances in experimental social psychology*, New York: Academic Press, Vol. 24, 1991.

Schwarz, N. , & Clore, G. L. , "Feelings and phenomenal experiences", in A. Kruglanski & E. T. Higgins (Eds.), *Social Psychology: Handbook of Basic Principles*, New York: Guilford, 2007.

Schwarz, N. , & Skurnik, I. , "Feeling and thinking implications or problem solving", in *The Nature of Problem Solving* (*Davidson, J. and Sternberg, R.* , eds), Cambridge University Press, 2003.

Sears, D. O. , "College sophomores in the laboratory: Influences of a narrow data base on social psychology's view of human nature", *Journal of Personality and Social Psychology*, No. 51, 1986.

Serbin, L. A. , Polishta, K. K. , & Gulko, J. , "The development of sex-typing in middle school", *Monographs of the Society for Research in Child Development*, Vol. 58, No. 2, 1993.

Sekaquaptewa, D. , Espinoza, P. , Thompson, M. , vargas, P. , & von Hippel, W. , "Stereotypic explanatory bias: Implicit stereotyping as a predictor of discrimination", *Journal of Experimental Social Psychology*, No. 39, 2003.

Sellers, R. M. , Rowley, S. A. J. , Chavous, T. M. , Shelton, J. N. , & Smith, M. , "Multidimensional inventory of black identity: Preliminary investigation of reliability and construct validity", *Journal of Personality and Social Psychology*, No. 73, 1997.

Shapiro, J. R., & Neuberg, S. L., "From stereotype threat to stereo-type threats: Implications of a multi-threat framework for causes, moderators, mediators, consequences, and interventions", *Personality and Social Psychology Review*, No. 11, 2007.

Sinclair, R. C., & Mark, M. M., "The influence of mood state on judgment and action: Effects on persuasion, categorization, social justice, person perception, and judgmental accuracy", in L. L. Martin & A. Tesser (Eds.), *The Construction of Social Judgments*, Hillsdale, NJ: Lawrence Erlbaum, 1992.

Siegel, E. F., Dougherty, M. R., & Huber, D. E., "Manipulating the role of cognitive control while taking the implicit association test", *Journal of Experimental Social Psychology*, No. 48, 2012.

Signorella, M. L., "Gender schemata: Individual differences and context effects", in L. S. Liben & M. L. Signorella (Eds.), *Children's Gender Schemata: New Directions for Child Development*, San Francisco, CA: Jossey-Bass, Vol. 38, 1987.

Signorella, M., Bigler, R., & Liben, L., "Developmental differences in children's gender schemata about others: A meta-analytic review", *Developmental Review*, No. 13, 1993.

Simpson, J. A., & Kenrick, D. T. (Eds.), *Evolutionary Social Psychology*, Mahwah, NJ: Lawrence Erlbaum Associates, 1997.

Sippola, L. K., Bukowski, W. M., & Noll, R. B., "Dimensions of liking and disliking underlying the same-sex preference in childhood ad early adolescence", *Merrill-Palmer Quarterly*, No. 43, 1997.

Slaby, R. G., & Frey, K. S., "Development of gender constancy and selective attention to same-sex models ", *Child Development*, No. 46, 1975.

Smith, E. R. , "The role of exemplars in social judgment ", in L. L. Martin & A. Tesser (Eds.), *The Construction of Social Judgment*, Hillsdale, NJ: Erlbaum, 1992.

Smith, C. T. , & Nosek, B. A. , "Affective focus increases the concordance between implicit and explicit attitudes", *Social Cognition*, No. 42, 2011.

Smith, E. R. , & Zárate, M. A. , "Exemplar and prototype use in social categorization", *Social Cognition*, Vol. 8, No. 3, 1990.

Smith, E. R. , & Zárate, M. A. , "Exemplar-based model of social judgment", *Psychological Review*, Vol. 99, No. 1, 1992.

Smuts, B. , "Male aggression against women: An evolutionary perspective", *Human Nature*, No. 3, 1992.

Smuts, B. , "The evolutionary origins of patriarchy", *Human Nature*, No. 6, 1995.

Son Hing, L. S. , Li, W. , & Zanna, M. P. , "Inducing hypocrisy to reduce prejudicial responses among aversive racists", *Journal of Experimental Social Psychology*, No. 38, 2002.

Spencer, S. J. , Steele, C. M. , & Quinn, D. M. , "Stereotype threat and women's math performance", *Journal of Experimental Social Psychology*, No. 35, 1999.

Srull, T. K. , & Wyer, R. S. , "Person memory and judgment", *Psychological Review*, No. 96, 1989.

Stangor, C. , & Duan, C. , "Effects of multiple tasks demands upon memory for information about social groups", *Journal of Experimental Social Psychology*, No. 27, 1991.

Stangor, C. , & Ford, T. E. , "Accuracy and expectancy-confirming o-rientations and the development of stereotypes and prejudice", *European Review of Social Psychology*, New York: Wiley, Vol. 3, 1992.

Stangor, C. , & Leary, S. , "Intergroup beliefs: Investigations from the social side ", *Advances in Experimental Social Psychology*, No. 38, 2006.

Stangor, C. , & Ruble, D. N. , " Differential influences of gender schemata and gender constancy on children's information processing and behavior", *Social Cognition*, No. 7, 1998.

Stangor, C. , Sullivan, L. A. , & Ford, T. E. , "Affective and cognitive determinants of prejudice", *Social Cognition*, No. 9, 1991.

Steele, C. M. , "A threat in the air: How stereotypes shape intellectual identity and performance", *American Psychologist*, No. 52, 1997.

Steele, C. M. , & Aronson, J. , "Stereotype threat and the intellectual performance of African Americans", *Journal of Personality and Social Psychology*, No. 69, 1995.

Steele, C. M. , Spencer, S. J. , & Aronson, J. , " Contending with group image: The psychology of stereotype and social identity threat", in M. P. Zanna (Ed.), *Advances in experimental social psychology*, San Diego: Academic Press, Vol. 34, 2002.

Stewart, T. L. , Amoss, R. T. , Weiner, B. A. , Elliott, L. A. , Parrott, D. J. , Peacock, C. M. , & Vanman, E. J. , " The psychophysiology of social action: Facial electromyographic responses to stigmatized groups predict antidiscrimination action", *Basic and Applied Social Psychology*, No. 35, 2013.

Stucke, T. S. , & Baumeister, R. F. , "Ego depletion and aggressive behaviour: Is the inhibition of aggression a limited resource?" *Europe-*

an *Journal of Social Psychology*, No. 36, 2006.

Stone, J. , "Battling doubt by avoiding practice: The effects of stereo-type threat on self-handicapping in White athletes", *Personality and Social Psychology Bulletin*, No. 28, 2002.

Stone, J. , Chalabaev, A. , & Harrison, C. K. , "The impact of stereotype threat on performance in sports", in M. Inzlicht & T. Schmader (Eds.), *Stereotype Threat: Theory, Process, and Application*, New York: Oxford University Press, 2011.

Sujoldzic, A. , Peternel, L. , Kulenovic, T. , & Terzic, R. , "Social determinants of health—a comparative study of Bosnian adolescents in different cultural contexts", *Collegium Antropologicum*, No. 30, 2006.

Taylor, T. R. , Williams, C. D. , Makambi, K. H. , Mouton, C. , Harrell, J. P. , Cozier, Y. , et al. , "Racial discrimination and breast cancer incidence in US black women: The black women's health study", *American Journal of Epidemiology*, No. 166, 2007.

Terrell, F. , Terrell, S. L. , & Miller, F. , "Level of cultural mistrust as a function of educational and occupational expectations among Black students", *Adolescence*, No. 28, 1993.

Todd, A. R. , Bodenhausen, G. V. , Richeson, J. A. , & Galinsky, A. D. , "Perspective taking combats automatic expressions of racial bias", *Journal of Personality and Social Psychology*, No. 100, 2011.

Trivers, R. L. , "Parental investment and sexual selection", in B. Campbell (Ed.), *Sexual Selection and the Descent of Man 1871 – 1971*, Chicago: Aldine, 1972.

Uhlmann, E. L. , Poelman, T. A. , & Nosek, B. , "Automatic associations: Personal attitudes or cultural knowledge?" in J. D. Hanson

（Ed.）, *Ideology, Psychology, and Law*, New York: Oxford University Press, 2002.

U. S. Department of Education, National Center for Educational Statistics, National Assessment of Educational Progress, 1990, 1992, 1996, 2002, and 2003 Mathematics Assessments, Retrieved March 10, 2005, from < http: //nces. ed. gov/ > : 2005.

Van de Vliert, E. , "Climato-economic origins of variation in ingroup favoritism", *Journal of Cross-Cultural Psychology*, Vol. 42, No. 3, 2011.

vanman, E. J. , Paul, B. Y. , Ito, T. A. , & Miller, N. , "The modern face of prejudice and the structural features that moderate the effect of cooperation on affect", *Journal of Personality and Social Psychology*, No. 73, 1997.

Vanman, E. J. , Saltz, J. L. , Nathan, L. R. , & Warren, J. A. , "Racial discrimination by low-prejudiced Whites: Facial movements as implicit measures of attitudes related to behavior", *Psychological Science*, No. 15, 2004.

Vargas, P. T. , Sekaquaptewa, D. , & von Hippel, W. , "Armed only with paper and pencil: 'Low-tech' measures of implicit attitudes", in B. Wittenbrink & N. Schwarz（Eds.）, *Implicit Measures of Attitudes*, New York: Guilford Press, 2007.

Vohs, K. D. , & Heatherton, T. F. , "Self-regulatory failure: A resource-depletion approach", *Psychological Science*, No. 11, 2000.

Von Hippel, W. , Sekaquaptewa, D. , & vargas, P. , "The linguistic intergroup bias as an implicit indicator of prejudice", *Journal of Experimental Social Psychology*, No. 33, 1997.

Wallaert, M. , Ward, A. , & Mann, T. , "Explicit control of implicit responses: Simple directives can alter IAT performance", *Social Psychology*, *No.* 41, 2010.

Watson. D., Clark, L. A., & Tellegen, A., "Development and validation of brief measures of positive and negative affect: The PANAS scales", *Journal of Personality and Social Psychology*, Vol. 54, No. 6, 1988.

Walton, G. M., & Cohen, G. L., "A question of belonging: Race, social fit, and achievement", *Journal of Personality and Social Psychology*, No. 92, 2007.

Watson, D., & Pennebaker, J. W., Health complaints, stress, and distress: Exploring the central role of negative affectivity. *Psychological Review*, No. 96, 1989.

Walton, G., Spencer, S. J., & Erman, S., "Affirmative meritocracy", *Manuscript under review*, 2011.

Wegener, D. T., Petty, R. E., & Smith, S. M., "Positive mood can increase or decrease message scrutiny: The hedonic contingency view of mood and message processing", *Journal of Personality and Social Psychology*, No. 69, 1995.

Weinraub, M., Clemens, L. P., Sockloff, A., Ethridge, T., Gracely, E., & Myers, B., "The development of sex role stereotypes in the third year: Relationships to gender labeling, gender identity, sex-typed toy preference, and family characteristics", *Child Development*, No. 55, 1984.

Westermann, R., Spies, K., & Stahl, G., "Relative Effectiveness and Validity of Mood Induction Procedures: A Meta-Analysis", *European Journal of Social Psychology*, Vol. 26, No. 4, 1996.

Williams, D. R., & Mohammed, S. A., "Discrimination and racial disparities in health: Evidence and needed research", *Journal of Behavioral Medicine*, No. 32, 2009.

Williams, D. R., "Race, socioeconomic status, and health: The added effect of racism and discrimination", in N. E. Adler & M. Marmot (Eds.), *Socioeconomic Status and Health in Industrial nations: Social, Psychological, and Biological Pathways*, New York: New York Academy of Sciences, Vol. 896, 1999.

Williams, D. R., & Rucker, T. D., "Understanding and addressing racial disparities in health care", H*ealth Care Financing Review*, Vol. 21, No. 4, 2000.

Williams, D. R., & Williams-Morris, R., "Racism and mental health: The African American experience", *Ethnicity and Health*, Vol. 5, No. 3 − 4, 2000.

Williams, J. E., & Bennett, S., "The definition of sex stereotypes via the adjective check list", *Sex Roles*, Vol. 1, No. 4, 1975.

Worth, L. T., & Mackie, D. M., "Cognitive mediation of positive affect in persuasion", *Social Cognition*, No. 5, 1987.

Wright, J., & Mischel, W., "Influence of affective on cognitive social learning person variables", *Journal of Personality and Social Psychology*, Vol. 43, No. 5, 1982.

Yen, I. H., Ragland, D. R., Grenier B. A., & Fisher, J. M., "Workplace discrimination and alcohol consumption: Findings from the San Francisco Muni health and safety study", *Ethnicity & Disease*, Vol. 9, No. 1, 1999.

Yeung, N. C. J., & von Hippel, C., "Stereotype threat increases the likelihood that female drivers in a simulator run over jaywalkers", *Accident Analysis & Prevention*, No. 40, 2008.

Yoshida, E. , Peach, J. M. , Zanna, M. P. , Spencer, S. J. , "Not all automatic associations are created equal: How implicit normative evaluations are distinct from implicit attitudes and uniquely predict meaningful behavior", *Journal of Experimental Social Psychology*, No. 48, 2012.

Zenasni, F. , & Lubart, T. I. , "Effects of emotional state on creativity", *Current Psychology Letters: Behavior, Brain and Cognition*, No. 2, 2002.

Ziv, T. , & Banaji, M. R. , "Representations of social groups in the early years of life", in S. T. Fiske and C. N. Macrae, *The SAGE Handbook of Social Cognition*, London: Sage, 2012.

Zuwerink, J. , Devine, P. , Monteith, J. , & Cook, D. , "Prejudice toward blacks: With and without compunction?" *Basic and Applied Social Psychology*, No. 18, 1996.